Algebra 1 Workbook

*The Self-Teaching Guide and Practice
Workbook with Exercises and Related
Explained Solution
You Will Get and Improve Your
Algebra 1 Skills and Knowledge
from A to Z*

by Richard Carter

Table of Contents

INTRODUCTION

Congratulations and thank you for purchasing Algebra 1 Workbook: The Self-Teaching Guide and Practice Workbook with Exercises and Related Explained Solution. You Will Get and Improve Your Algebra 1 Skills and Knowledge from A to Z.

Algebra is a very noteworthy subfield of mathematics in its versatility alone if nothing else. The most general and the most commonly used definition of algebra involves the studying of certain mathematical symbols as well as the study of the manipulation of these symbols. Mathematical symbols are one of the most basic elements of mathematics, aside from numbers themselves and operation symbols, so the study of these symbols is one of the most important studies that one can take up as far as mathematics is concerned.

Within this book, you will find some of the most important topics regarding algebra.

These are including but are never limited to the following: understanding integers and basic operations; inequalities and one step operations; fractions and factors; the main rules of arithmetic; linear equations in the coordinate plane, functions, expressions, equation, real numbers; solving linear equations; visualizing functions that are linear, equations that are linear, inequalities that are linear, linear equations and inequalities (systems of); functions that have exponents and exponential functions, polynomials; equations with quadratics; expressions with radicals; equations with radicals expressions that are rational; and finally, intermediate topics in algebra.

Note: This book has been reviewed taking into account the comments and suggestions provided by the readers in their reviewes.

In order to get the best result by this workbook, the reader/student should be a good knowledge of math basis, in any case the workbook provides at the beginning, a short refresh of math basis needed to apply in Algebra 1 exercises.

The book is so structured:

PART 1

TOPIC TITLE:

- **How to solve:** the book explains step by step how to solve each operation, including formula, rules etc.

- **Problem Solved**

  In this section there are some problems (exactly some of the problems proposed in the next chapter) step by step solved and explained in order to provide all the needed info for the different cases, so the student should be able to develop and solve the proposed problems in the next chapter "Problems".

- **Problems**: The workbook proposes different problems to solve regarding that topic, with different levels of difficulty.

PART 2 SOLUTIONS

It provides the solutions for all the problems for each TOPIC of *PART* 1.

REFRESH SOME MATH BASIS

Divisibility Rules

How to Solve

Remember the most used divisibility rules:

A number is divisible:

by 2	if the last digit is 0,2,4,6,8
by 3	If the sum of the digits is a number divisible by 3
by 5	if the last digit is 0 or 5
by 10	if the last digit is 0

Problems

1)   Indicate Yes or NO if the number is divisible by 2

Number	divisible by 2, Yes or No
53,764	
1,246	
69,749	
738	
9,350	
345	
4,348	
15	
92,576	
3,273	

2)   Indicate Yes or NO if the number is divisible by 3

Number	divisible by 3, Yes or No
1,353	
36,696	
4,567	
35	
8,241	
56	
9,132	
25,788	
7,901	
99	

3) Indicate Yes or NO if the number is divisible by 5

Number	divisible by 5, Yes or No
75	
450	
4,658	
81,270	
3,825	
6,200	
12,835	
97	
1,011	
52,170	

4) Indicate Yes or NO if the number is divisible by 10

Number	divisible by 10, Yes or No
100	
3,568	
40,375	
785,420	
58,743	
736,271	
52,340	
4,890	
34,719	
89,770	

Least Common Multiple (LCM)

The least common multiple (or the LCM) is our smallest multiple that exists between two or more numbers.
This will usually be larger than both numbers in question.

For example, 15 is the least common multiple between 3 and 5, because: $3 \times 5 = 15$ and $5 \times 3 = 15$

This is our smallest possible multiple which the two numbers have in common.

To determine the LCM of any two given numbers, you must first find multiples of both until you come across the smallest multiple that both have in common.

There is virtually no shorthand method of doing this, so practice is necessary to master the skill.

In case you have to add or subtract fractions with different denominators, you have to find the LCM.

To do that you have to write the prime factorization for each denominator.

How to Solve

Example: Find the LCM for 8 and 12

1)   Prime Factorization of 8:

$$8 \; is \; divisible \; by \; 2$$
$$8 \mid 2$$
$$4 \; and \; is \; divisible \; by \; 2$$
$$4 \mid 2$$
$$2 \; \; is \; divisible \; by \; 2$$
$$2 \mid 2$$
$$1 \; is \; a \; prime \; number, so \; the \; prime \; factorization \; of \; 8 \; is.$$

$$\mathbf{8 = 2 \times 2 \times 2}$$

2)   Prime Factorization of 12

$$12 \; and \; is \; divisible \; by \; 3$$
$$12 \mid 3$$
$$4 \; \; is \; divisible \; by \; 2$$
$$4 \mid 2$$
$$2 \; and \; is \; divisible \; by \; 2$$
$$2 \mid 2$$
$$1 \; is \; a \; prime \; number, so \; the \; prime \; factorization \; of \; 12 \; is:$$

$$\mathbf{12 = 3 \cdot 2 \cdot 2}$$

Now take a look to both numbers 8 and 12:

- The most the factor 2 appears is three times
- The most the factor 3 appears is once

So, to find the LCM between 8 and 12 you must multiply

$$LCM = 2 \cdot 2 \cdot 2 \cdot 3 = 24$$

Quick method

- *multiples of 8 are*: $8, 16, \mathbf{24}$
- *multiples of 12 are*: $12, \mathbf{24}$

The Least Common Multiple between 8 and 12 is 24 because it is the lowest common multiples for 8 and 12.

Problems Solved

Find the LCM for the following couples of numbers:

The following problems are taken by the next chapter.

1) **Problem 2**: LCM between 5 and 6:

$$5: 5,10,15,20,25, \mathbf{30}$$
$$6: 6,12,18,24, \mathbf{30}$$

The Least Common Multiple **LCM** $=$ **30**

2) Problem 7: LCM between 11 *and* 2:

$$\mathbf{11}: \ 11, \mathbf{22}$$
$$2: 2,4,6,8,10,12,14,16,18,20, \mathbf{22}$$

The Least Common Multiple **LCM** $=$ **22**

3) Problem 8: LCM between 8 *and* 12:

$$\mathbf{8}: \ 8,16, \mathbf{24}$$
$$\mathbf{12}: \ 12, \mathbf{24}$$

The Least Common Multiple **LCM** $=$ **24**

Problems

Find the LCM for the following couples of numbers

1	3, 7
2	5, 6
3	12, 20
4	4, 6
5	4, 7
6	12, 5
7	11, 2
8	8, 12

Simplify Fractions

How to Solve

Simplify with Criss Cross method

Whenever is it possible apply the divisibility rules with criss cross method.

Example 1: 3 and 6 are both divisible by 3 and 5 and 15 are both divisible by 5, so

$$\frac{3}{15} \times \frac{5}{6} = \frac{3 \div 3}{15 \div 5} \times \frac{5 \div 5}{6 \div 3} = \frac{1}{3} \times \frac{1}{2} = \frac{1}{6}$$

Example 2: 6 and 14 are both divisible by 2, so

$$\frac{6}{13} \times \frac{1}{14} = \frac{6 \div 2}{13} \times \frac{1}{14 \div 2} = \frac{3}{13} \times \frac{1}{7} = \frac{3}{91}$$

Example 3: 20 and 5 are both divisible by 5, so:

$$20 \times \frac{2}{5} = 20 \div 5 \times \frac{2}{5 \div 5} = 4 \times 2 = 8$$

Simplify between Numerator and Denominator

Whenever is it possible, simplify numerator and denominator by applying the divisibility rules.

Example 1:

$$\frac{25}{5} = \frac{25 \div 5}{5 \div 5} = 5$$

Example 2:

$$\frac{20}{12} = \frac{20 \div 2}{12 \div 2} = \frac{10}{6} = \frac{10 \div 2}{6 \div 2} = \frac{5}{3}$$

Problems Solved

1) **Problem 2**: Simplify the fraction

$$\frac{8}{12}$$

8 and 12 can be divisible by 2, so

$$\frac{8}{12} = \frac{8 \div 2}{12 \div 2} = \frac{4}{6}$$

4 and 6 can be divisible by 2 again, so

$$\frac{4 \div 2}{6 \div 2} = \frac{2}{3}$$

Solution

$$\frac{2}{3}$$

2) **Problem 5**: Simplify the fraction of this multiplication

$$\frac{4}{12} \times \frac{1}{8}$$

we can do two different simplifications:

1) Criss Cross between 4 and 8, both divisible by 4

2) Numerator and denominator 4 and 12 both dividible by 4

1) criss cross:

$$\frac{4}{12} \times \frac{1}{8} = \frac{4 \div 4}{12} \times \frac{1}{8 \div 4} = \frac{1}{12} \times \frac{1}{2} = \frac{1}{24}$$

2) numerator and denominator:

$$\frac{4}{12} \times \frac{1}{8} = \frac{4 \div 4}{12 \div 4} \times \frac{1}{8} = \frac{1}{3} \times \frac{1}{8} = \frac{1}{24}$$

Problems

1	$\dfrac{15}{5}$
2	$\dfrac{8}{12}$
3	$\dfrac{1}{4} \times 6$
4	$\dfrac{3}{6} \times \dfrac{6}{9}$
5	$\dfrac{4}{12} \times \dfrac{1}{8}$
6	$\dfrac{18}{12}$
7	$\dfrac{25}{5}$
8	$\dfrac{4}{3} \times \dfrac{9}{2}$

Operations with Fractions

How to Solve

1) Addition: same denominators

$$\frac{A}{B} + \frac{C}{B} = \frac{A+C}{B}$$

Example:

$$\frac{2}{3} + \frac{5}{3} = \frac{7}{3}$$

2) Addition: different denominators

When the denominators are different, you must find the LCM between all the denominators, so

$$\frac{A}{B} + \frac{C}{D} = \frac{A \times (LCM \div B) + C \times (LCM \div D)}{LCM}$$

Example:

$$\frac{5}{2} + \frac{3}{5}$$

find the LCM between 2 and 5
2: 2,4,6,8, **10**
5: 5, **10**
$\boldsymbol{LCM = 10}$

$$\frac{5}{2} + \frac{3}{5} = \frac{5 \times (10 \div 2) + 3 \times (10 \div 5)}{10} = \frac{(5 \times 5) + (3 \times 2)}{10}$$

$$= \frac{25 + 6}{10} = \frac{\mathbf{31}}{\mathbf{10}}$$

Subtraction: same denominators

$$\frac{A}{B} - \frac{C}{B} = \frac{A - C}{B}$$

Example:

$$\frac{7}{3} - \frac{5}{3} = \frac{7 - 5}{3} = \frac{2}{3}$$

3)   Subtraction: different denominators

$$\frac{A}{B} - \frac{C}{D} = \frac{A \times (LCM \div B) - C \times (LCM \div D)}{LCM}$$

Example:

$$\frac{5}{8} - \frac{3}{12}$$

find the LCM between 8 *and* 12

8: 8, 16, **24**

12: 12, **24**

LCM = 24

$$\frac{5}{8} - \frac{3}{12} = \frac{5 \times (24 \div 8) - 3 \times (24 \div 12)}{24} =$$

$$\frac{(5 \times 3) - (3 \times 2)}{24} = \frac{15 - 6}{24} = \frac{9}{24} = \frac{9 \div 3}{24 \div 3} = \frac{3}{8}$$

4) Division:

$$\frac{A}{B} \div \frac{C}{D} = \frac{A}{B} \cdot \frac{D}{C} = \frac{AD}{BC}$$

Example:

$$\frac{6}{5} \div \frac{3}{20} =$$

the division becomes multiplication, so

multiply the first fraction by the reverse of second fraction

$$\frac{6}{5} \times \frac{20}{3} =$$

if you look well, you see that first we can do

the simplification with criss cross, so

$$\frac{6 \div 3}{5 \div 5} \times \frac{20 \div 5}{3 \div 3} = 2 \times 4 = 8$$

Problems

1	$\dfrac{2}{7} + \dfrac{5}{2}$
2	$\dfrac{6}{5} \div \dfrac{3}{20}$
3	$\dfrac{30}{4}$
4	$\dfrac{7}{8} - \dfrac{3}{4}$
5	$\dfrac{7}{5} \times \dfrac{20}{21}$
6	$\dfrac{2}{5} \div \dfrac{3}{15}$

7	$\dfrac{9}{3} - \dfrac{3}{12}$
8	$\dfrac{8}{5} + \dfrac{3}{4}$
9	$\dfrac{40}{12}$
10	$\dfrac{2}{3} \div \dfrac{3}{15}$
11	$\dfrac{90}{10}$

INTEGERS

Absolute Value

The absolute value is defined as the magnitude of a number without regard to its sign. In other words, exactly how far a number is away from zero, with no attention paid as to whether the number is positively or negatively signed. Absolute value only determines where a certain number is featured on a number line, it does not imply a real value to the number though.

One example of absolute value differing from the real value of an actual number is -6. The absolute value, in this case, would be 6, since -6 is 6 units away from zero. While -6 is less than zero, its absolute value remains more than zero. 6 also would have the absolute value 6, since it is placed 6 units to the right of zero. The only difference, in this case, is the direction at which the number diverts from zero on a number line.

Once you have determined the absolute value of your number, you must then place "|" marks around the value to indicate that it is an absolute value.

To determine a negative number's absolute value, you always must remember to remove the negative from the number in order to arrive at your final answer.

How to write

1) Write the absolute value of -9, 3 and -17

   |9|, |3|, |17|

2) Write the absolute value of 2 and -14

   |2|, |14|

Adding and Subtracting

We should at first try to think of adding in subtracting integers in terms of their placement on the number line. When looking at it this way, we must add positive integers by moving by the value added to the right in the number line. In order to add negative integers, we must move by the value added to the left in it. In order to subtract from either positive or negative integers, we must go to the opposite directions respectively the values being subtracted.

How to Solve

Below is shown the number line. On the right of 0, there are the positive numbers, on the left of 0 there are the negative numbers.

- In order to add integers that have the same sign, we must keep the sign and add by the absolute values of the integers.

Example : $2 + 6 = 8$

- In order to add integers that have different signs, we must keep the sign that belongs to the integer with the largest absolute value and then subtracts it by our integer that has the smallest absolute value from it.

  Examples:

$$-5 + 9 = +4$$
$$+12 - 5 = +7$$
$$-6 + 1 = -5$$

- **Integers Opposite Property**: In order to subtract an integer, it is necessary to add the integer's opposite.

  Examples:

$$the\ opposite\ of\ -9 = 9$$
$$the\ opposite\ of\ 11\ = -11$$

- **Double Negative**: One final thing to keep in mind when solving for integers is that double negatives equal positives, so the opposite positive number is always to be found when we come across **double negatives**:

$$-(-a) = +a$$

Example:

$$-(-6) = +6$$

Let's now solve for some certain examples **using numbers line**:

1) to solve for 2 + 6, we would start at 2 on our number line and then move to the right 6 units, which would, in turn, make us arrive at 8, so 2 + 6 = 8

2) To solve for -2 + 5, we would need to start at -2 and then move to the right 5 units, which would bring us to 3, so -2 + 5 = 3

3) To solve for 3 + (-7), we would actually have to eliminate the plus sign and only use the minus from -7. Our equation would then look like 3 - 7 and to solve it, we would start out at 3 and move to the left on our number line by 7 units, bringing us to -4, so:

$$3 + (-7) = -4$$

4) In order to solve for - 4 + (-1), we would need to do the same; to replace the plus sign with the minus from the -7. Our equation would now look like - 4 - 1. We would then start at - 4 and move to the left by 1 unit, which would give us -5, so:

$$-4 + (-1) = -5$$

Remember:

1) From this point on, the positive numbers will be indicated without the sign +, for example +7 will be write as 7.

2) Regarding the multiplication, from this point on, we will use the sign $\cdot$ (dot) instead of the $\times$. When the multiplication involves parenthesis, the multiplication doesn' t use any sign, so for example

$$2(3) \; means \; 2 \times 3,$$

3) A sign + before a parenthesis doesn't change the sign inside it.

Example:

$$6 + (-2) = 6 - 2 = 4$$

4) A sign *minus* $-$ before a parenthesis changes the signs inside it.

Examples:

1) $4 - (1) = 4 - 1 = 3$

2) $5 - (-7) = 5 + 7 = 12$

3) $4 - (-3 + 4 - 7) = 4 + 3 - 4 + 7 = 10$

Problems Solved

Below you find the complete solutions regarding some of problems the reader should do in the next chapter "Problems".

In this way the student should be able to do the remaining problems and check the correct solution in the chapter solution.

1) Problem 4:

$$-5 - -5 = -5 + 5 = 0$$

double negative becomes positive, in other

words the negative sign, changes the next signs, so

$$-5 + 5 = 0$$

2) Problem 10:

$$-1 + -4 =$$

the positive sign, doesn't change the next sign, so

$$-1 - 4 = -5$$

3) Problem 20:

$$-\frac{13}{4} - \frac{13}{7}$$

$$\text{LCM} = 28$$

$$-\frac{13}{4} - \frac{13}{7} = \frac{-13 \cdot (28 \div 4) - 13 \cdot (28 \div 7)}{28} =$$

$$\frac{-13(7) - 13(4)}{28} = \frac{-91 - 52}{28} = -\frac{\mathbf{143}}{\mathbf{28}}$$

4) Problem 11

$$-\frac{6}{5} - -\frac{7}{4}$$

$$LCM = 20$$

$$-\frac{6}{5} - -\frac{7}{4} = \frac{-6(20 \div 5) - -7(20 \div 4)}{20} =$$

$$\frac{-6(4) - -7(5)}{20} = \frac{-24 - -35}{20} = \frac{-24 + 35}{20} = \frac{\mathbf{11}}{\mathbf{20}}$$

5) Problem 13

$$-\frac{1}{3} + -\frac{23}{8} =$$

$$LCM = 24$$

$$\frac{-1 \cdot (24 \div 3) - 23(24 \div 8)}{24} = \frac{-1 \cdot (8) - 23 \cdot (3)}{24} =$$

$$\frac{-8 - 69}{24} = -\frac{77}{24}$$

6) Problem 2

$$-7 - -2 = -7 + 2 = -5$$

Problems

1	$-3 - 2$	**11**	$-\dfrac{6}{5} - -\dfrac{7}{4}$
2	$-7 - -2$	**12**	$-\dfrac{17}{7} + -\dfrac{10}{3}$
3	$8 - -1$	**13**	$-\dfrac{1}{3} + -\dfrac{23}{8}$
4	$-5 + 5$	**14**	$-\dfrac{11}{6} - \dfrac{3}{2}$
5	$6 - -3$	**15**	$\dfrac{9}{2} - \dfrac{15}{8}$
6	$-2 + 1$	**16**	$-\dfrac{8}{5} - \dfrac{11}{5}$
7	$-6 - 3$	**17**	$\dfrac{5}{2} + -\dfrac{4}{3}$
8	$5 - 4$	**18**	$-\dfrac{23}{7} + -\dfrac{7}{4}$
9	$-5 + 3$	**19**	$\dfrac{22}{7} - \dfrac{11}{8}$
10	$-1 + -4$	**20**	$-\dfrac{13}{4} - \dfrac{13}{7}$

Multiplying

$$(+a) \cdot (+b) = +(a \cdot b)$$
$$(-a) \cdot (+b) = -(a \cdot b)$$
$$(+a) \cdot (-b) = -(a \cdot b)$$
$$(-a) \cdot (-b) = +(a \cdot b)$$

How to Solve

1) $(+3) \cdot (+5) = +(15) = 15$

2) $(-3) \cdot (+4) = -(12) = -12$

3) $(+6) \cdot (-4) = -(24) = -24$

4) $(-2) \cdot (-9) = +(18) = 18$

5)

$$-\frac{35}{6} \cdot -\frac{8}{9} = -\frac{35}{3} \cdot -\frac{4}{9} = \frac{140}{27}$$

Remember to simplify fractions applying the divisibility rules between numerator and denominator of the same fraction or between numerator and denominator and vice-versa (criss-cross) in case of multiplication of fractions. In this example the denominator 6 of first fraction with the numerator 8 of second fraction.

$$-\frac{35}{6} \cdot -\frac{8}{9} = -\frac{35}{3} \cdot -\frac{4}{9} = \frac{140}{27}$$

Problems Solved

Below you find the complete solutions regarding some of problems the reader should do in the next chapter "Problems".

1) Problem 1:

$$3 \cdot -1 \cdot 8 =$$

First decide the final sign, so

$$+ \cdot - \cdot + = -$$

the final sign is $-$, then evaluate the product, so:

$$-(3 \cdot 1 \cdot 8) = -24$$

2) Problem 5:

$$-2 \cdot -2 \cdot 2 = +(2 \cdot 2 \cdot 2) = 8$$

3) Problem 11:

$$7 \cdot -\frac{4}{5} = -\left(7 \cdot \frac{4}{5}\right) = -\frac{7 \cdot 4}{5} = -\frac{28}{5}$$

4) Problem 13:

$$-\frac{35}{6} \cdot -\frac{8}{9} = +\left(\frac{35}{6} \cdot \frac{8}{9}\right) =$$

now, we can simplify with criss $-$ cross, so

$$\frac{35}{6 \div 2} \cdot \frac{8 \div 2}{9} = \frac{35}{3} \cdot \frac{4}{9} = \frac{35 \cdot 4}{3 \cdot 9} = \frac{140}{27}$$

Problems

1	$3 \cdot -1 \cdot 8$	**11**	$7 \cdot -\dfrac{4}{5}$
2	$3 \cdot -6 \cdot -8$	**12**	$\dfrac{2}{3} \cdot -\dfrac{3}{2}$
3	$-5 \cdot 7 \cdot -10$	**13**	$-\dfrac{35}{6} \cdot -\dfrac{8}{9}$
4	$2 \cdot -9 \cdot 3$	**14**	$\dfrac{13}{10} \cdot -\dfrac{1}{3}$
5	$-2 \cdot -2 \cdot 2$	**15**	$-5 \cdot -\dfrac{1}{2}$
6	$-10 \cdot 7 \cdot -7$	**16**	$-\dfrac{19}{9} \cdot \dfrac{7}{5}$
7	$-6 \cdot -7 - \cdot 3$	**17**	$\dfrac{1}{4} \cdot -\dfrac{19}{10}$
8	$10 \cdot -3 \cdot -8$	**18**	$-\dfrac{11}{6} \cdot \dfrac{4}{5}$
9	$4 \cdot 6 \cdot -2$	**19**	$-\dfrac{7}{4} \cdot \dfrac{5}{3}$
10	$2 \cdot 7 \cdot -10$	**20**	$\dfrac{8}{5} \cdot -\dfrac{17}{10}$

Dividing

$$\frac{-a}{-b} = +\frac{a}{b}$$

$$\frac{-a}{b} = -\frac{a}{b}$$

$$\frac{a}{-b} = -\frac{a}{b}$$

How to Solve

Example 1:

$$\frac{-2}{-6}$$

apply the rule $\frac{-a}{-b} = +\frac{a}{b}$, *and then simplify the result, so:*

$$\frac{-2}{-6} = \frac{2}{6} = \frac{2 \div 2}{6 \div 2} = \frac{1}{3}$$

Example 2:

$$\frac{-8}{4}$$

apply the rule $\frac{-a}{b} = -\frac{a}{b}$ *and then simplify the result, so*

$$\frac{-8}{4} = -\frac{8}{4} = -\frac{8 \div 4}{4 \div 4} = -2$$

Example 3:

$$\frac{18}{-2}$$

apply the rule $\dfrac{a}{-b} = -\dfrac{a}{b}$ *and then simplify the result, so*

$$\frac{18}{-2} = -\frac{18}{2} = -\frac{18 \div 2}{2 \div 2} = -9$$

Example 4:

To make the division between fractions, you have to change the division in a mutìltiplicatio. To do this, you multiply the fisrst fraction by the reverse of the second fraction.

$$\frac{\frac{7}{2}}{-\frac{31}{8}} = \frac{7}{2} \cdot -\frac{8}{31} = \frac{7}{2 \div 2} \cdot -\frac{8 \div 2}{31} =$$

$$= \frac{7}{1} \cdot -\frac{4}{31} = -\frac{28}{31}$$

Example 5:

To make the division between fractions, you have to change the division in a mutìltiplicatio. To do this, you multiply the fisrst fraction by the reverse of the second fraction.

$$\frac{-\frac{3}{2}}{-\frac{11}{8}} = -\frac{3}{2} \cdot -\frac{8}{11} = \frac{24}{22}$$

Problems Solved

1) Problem 7:

$$\frac{-64}{-8} =$$

$$\frac{64}{8} = \frac{64 \div 8}{8 \div 8} = 8$$

2) Problem 2:

$$\frac{36}{4} = \frac{36 \div 4}{4 \div 4} = 9$$

3) Problem 15:

$$\frac{\dfrac{12}{7}}{\dfrac{8}{5}} =$$

$$\frac{12}{7} \cdot \frac{5}{8} = \frac{60}{56} = \frac{60 \div 2}{56 \div 2} = \frac{30}{28} = \frac{30 \div 2}{28 \div 2} = \frac{15}{14}$$

because we have simplified numerator

and denominator twice

Problems

1	$-\dfrac{16}{8}$	**10**	$\dfrac{\frac{45}{8}}{\frac{1}{6}}$
2	$\dfrac{36}{4}$	**11**	$\dfrac{\frac{14}{9}}{\frac{1}{3}}$
3	$-\dfrac{45}{9}$	**12**	$\dfrac{\frac{13}{3}}{-\frac{5}{6}}$
4	$-\dfrac{15}{5}$	**13**	$\dfrac{-2}{-\frac{18}{5}}$
5	$\dfrac{60}{30}$	**14**	$\dfrac{-\frac{17}{7}}{\frac{4}{5}}$
6	$-\dfrac{16}{2}$	**15**	$\dfrac{\frac{12}{7}}{\frac{8}{5}}$
7	$\dfrac{-64}{-8}$	**16**	$\dfrac{\frac{7}{2}}{-\frac{31}{8}}$
8	$-\dfrac{70}{10}$	**17**	$\dfrac{-\frac{3}{2}}{-\frac{11}{8}}$
9	$\dfrac{-100}{-10}$	**18**	$\dfrac{1}{-\frac{8}{5}}$

POWERS AND EXPONENTS

Now we come to powers and exponents. A power is defined as being an expression that indicates what to multiply by the same exact factor as the base number. The base number here is the one being multiplied by itself. Here are some example of base numbers and their powers:

The exponent is the small number superscript above its base number. This number includes the power that the base number is to be multiplied by, i.e. how many times it is necessary to multiply the base by itself.

Properties of exponents Overview

1) $a^0 = 1$ $ex. \, 3^0 = 1$

2) $a^{-n} = \frac{1}{a^n}$ $ex. \, 5^{-2} = \frac{1}{5^2} = \frac{1}{25}$

3) $\frac{1}{a^{-n}} = a^n$ $ex. \, \frac{1}{2^{-3}} = 2^3$

4) $a^m \cdot a^n = a^{m+n}$ $ex. \, 4^{-9} \cdot 4^4 = 4^{-9+4} = 4^{-5} = \frac{1}{4^5}$

5) $(a^m)^{\,n} = a^{m \cdot n}$ $ex. \, (2^7)^{\,3} = 2^{7 \cdot 3} = 2^{21}$

6) $(ab)^n = a^n \cdot b^n$ $ex. \, (2 \cdot 3)^{-4} = 2^{-4} \cdot 3^{-4}$

7) $\left(\frac{a}{b}\right)^n \textbf{ with } b \neq 0 = \frac{a^n}{b^n}$ $ex. \, \left(\frac{2}{3}\right)^8 = \frac{2^8}{3^8}$

Example 1:

$$-2^4 = -2 \cdot 2 \cdot 2 \cdot 2 = -16$$

Example 2:

$$3^0 = 1$$

When the exponent is 0, the result is always 1 for all the bases of the power $\neq 0$.

Example 3:

$$5^{-2} = \frac{1}{5^2} = \frac{1}{25}$$

Example 3:

$$-4^{-3} = \frac{1}{(-4)^3} = \frac{1}{-64} = -\frac{1}{64}$$

Example 4:

$$\left(a^{-3}\right)^{-2} = \frac{1}{\left(a^{-3}\right)^2} = \frac{1}{a^{-6}} = a^6$$

Product property

$$x^a \cdot x^b = x^{a+b}$$

How to Solve

1) $3^3 \cdot 3^2 = 3^5 = 243$

2) $2^{-2} \cdot 2^5 = 2^{-2+5} = 2^3 = 8$

3) $2^{-2} \cdot 2^5 = 2^{-2+5} = 2^3 = 8$

4) $4n^4 \cdot n \cdot 4n^2 = (4 \cdot 1 \cdot 4)n^{4+1+2} = 16n^7$

5) $3x^0 \cdot 2x^3 = ((3 \cdot 1) \cdot 2)\, x^3 = 6x^3$

   <u>Remember</u>: when the exponent is 0 the result is always 1
   So $x^0 = 1$

6) $4v^0 \cdot v \cdot 2v^0 = (4 \cdot 1 \cdot 2)v^1 = 8v$

7) $4x^3 \cdot 3x^{-1} = (4 \cdot 3)x^{\,3-1} = 12x^2$

8) $4n^{-3} \cdot 3n^2 = 12n^{-1} = \frac{12}{n^1} = \frac{12}{n}$

9) $3m^{-2}m^4 = 3m^2$

10) $\quad v^0 3v^2 = 1 \cdot 3v = 3v$

11) $\quad 4n^{-2} \cdot 3n^{-4} \cdot 4n^4 = 48n^{-2-4\,+4} = 48n^{-2} = \frac{48}{n^2}$

Problems

1	$4m \cdot 2m^{-1}$	**11**	$2x^4 \cdot x^2$
2	$4v^0 \cdot v \cdot 2v^0$	**12**	$3x^0 \cdot 2x^3$
3	$4a^{-2} \cdot 2a^3$	**13**	$v^0 \cdot 3v^2$
4	$4n^4 \cdot n \cdot 4n^2$	**14**	$4n^{-2} \cdot 3n^{-4} \cdot 4n^4$
5	$4n^{-3} \cdot 3n^2$	**15**	$4n^{-3} \cdot 3n^0 \cdot 4n^{-4}$
6	$2v \cdot 3v^4$	**16**	$x^4 \cdot 3x^4$
7	$m^{-4} \cdot m^4$	**17**	$3p \cdot 3p^2 \cdot 4p^{-4}$
8	$3n^{-3} \cdot n^4$	**18**	$3x \cdot x^{-3}$
9	$3m^{-2} \cdot m^4$	**19**	$3k^2 \cdot 4k$
10	$4x^3 \cdot 3x^{-1}$	**20**	$4k \cdot 4k^3$

Power Property

$$(x^a)^b = x^{a \cdot b}$$

How to Solve

1) $(2^3)^2 = 2^{3 \cdot 2} = 2^6$

2) $(2x^3)^4 = 2^4 x^{3 \cdot 4} = 16x^{12}$

3) $(-x^4)^{-3} = -x^{-12} = -\dfrac{1}{x^{12}}$

4) $(5n^{-5})^3 = 5^3 n^{-5 \cdot 3} = 125n^{-15} = \dfrac{125}{n^{15}}$

5) $(-3v^3)^4 = (-3)^4 v^{3.4} = 81^{12}$

6) $(5n^{-1})^0 = 5^0 n^{-1 \cdot 0} = 1 \cdot n^0 = 1 \cdot 1 = 1$

Problems

1	$(3x^2)^3$	**11**	$(3x^2)^3$
2	$(v^3)^3$	**12**	$(-3v^3)^4$
3	$(3x^3)^0$	**13**	x^5
4	$(x^3)^2$	**14**	$(-x^4)^{-3}$
5	$(2b^2)^3$	**15**	$(4x)^2$
6	$(3m^3)^2$	**16**	$(2p^{-5})^{-3}$
7	$(3x)^3$	**17**	$(5n^{-1})^0$
8	$(3k)^2$	**18**	$(5n^{-5})^3$
9	$(3b^3)^3$	**19**	$(-2x^3)^4$
10	$(2m^2)^2$	**20**	$(-5m^3)^2$

Quotient Property

$$x^a \div x^b = x^{a-b}$$

If the numerator and denominator have different basis, you can simplify the fraction by dividing both by the minimum common factor

How to Solve

Example 1: this is the problem 4 of the next chapter

$$\frac{x^3}{3x}$$

now, divide both by x, so

$$\frac{\frac{x^3}{x}}{\frac{3x}{x}} = \frac{x^3}{x} \cdot \frac{x}{3x} = \frac{x^2}{3}$$

Example 2:

$$2^8 \div 2^5 = 2^{8-5} = 2^3 = 8 \quad \text{(since same basis)}$$

Example 3:

$$\frac{3^8}{3^5} = 3^{8-5} = 3^3 = 27 \qquad \text{(since same basis)}$$

Example 4: this is the problem 4 of the next chapter

$$\frac{x}{3r^2} = \frac{1}{3r}$$

Example 5: this is the problem 14 of the next chapter

$$\frac{-5x^{-3}}{-2x^{-3}}$$

These are different basis, but you can simplify the fraction by dividing numerator and denominator by x^{-3}, so

$$\frac{-5x^{-3}}{-2x^{-3}} =$$

$$\frac{\frac{-5x^{-3}}{x^{-3}}}{\frac{-2x^{-3}}{x^{-3}}} = \frac{-5x^{-3}}{x^{-3}} \cdot \frac{x^{-3}}{-2x^{-3}} = \frac{-5}{-2} = \frac{5}{2}$$

Example 6: this is the problem 20 of the next chapter

$$\frac{3p^{-5}}{3p^{-2}} = divide\ by\ 3 = \frac{\frac{3p^{-5}}{3}}{\frac{3p^{-2}}{3}} = \frac{p^{-5}}{p^{-2}} = p^{-5--2}$$

$$= p^{-3} = \frac{1}{p^3}$$

Problems

1	$\dfrac{m^3}{m^0}$	**11**	$-\dfrac{2r^5}{4r^5}$
2	$\dfrac{r}{3r^2}$	**12**	$\dfrac{2n^4}{-4n}$
3	$\dfrac{2k^2}{3k^3}$	**13**	$\dfrac{5k^{-3}}{k^5}$
4	$\dfrac{x^3}{3x}$	**14**	$\dfrac{-5x^{-3}}{-2x^{-3}}$
5	$\dfrac{2x^2}{x^3}$	**15**	$\dfrac{-3n^{-1}}{-4n^2}$
6	$\dfrac{2n^0}{2n^2}$	**16**	$\dfrac{2n^{-5}}{n^3}$
7	$\dfrac{3x^0}{3x}$	**17**	$\dfrac{4x^2}{5x}$
8	$\dfrac{2p^0}{3p}$	**18**	$-\dfrac{4m^3}{5m^0}$
9	$\dfrac{2a^2}{a^0}$	**19**	$\dfrac{-5r^4}{-2r}$
10	$\dfrac{2n^2}{3n^2}$	**20**	$\dfrac{3p^{-5}}{3p^{-2}}$

ALGEBRIC OPERATIONS IN THE CORRECT ORDER

The first topic we should discuss regarding algebra is solving the algebraic equations. Algebraic equations are any one of the traditional operations of arithmetic. These are the most basic math skills that we learn in elementary school, being addition, subtraction, multiplication, and division. There is an order that we need to follow in solving algebraic equations. This is what is known as **PEMDAS** (<u>Parentheses, Exponents, Multiplication, Division, Addition, and Subtraction</u>). We need to observe this rule whenever we are faced with an algebraic equation.

How to Solve

Example 1:

$$x = 2 + 3(8)$$

we would multiply $3 \cdot 8$ before adding 2 to it because multiplication come before addition in PEMDAS, so:

$$x = 2 + 24 = 26$$

Example 2:

$$x = 4(3 + 2) - 1$$

$$4(5) - 1 = 20 - 1 =$$

$$x = 19$$

Example 3:

$$2 + (3 - 1) \cdot 3^2 = 2 + 2 \cdot 3^2 =$$

$$2 + (2) \cdot 9 = 2 + 18 = 20$$

Example 4:

$$10 - 5 \times 2^2 + (36 \div 6) - 3 =$$

$$10 - 5 \times 2^2 + (6) - 3 =$$

$$10 - 5 \times 4 + 6 - 3 =$$

$$10 - 20 + 6 - 3 =$$

$$10 - 14 - 3 = -7$$

Example 5:

$$\frac{13}{6} - \frac{5}{2} \cdot \frac{1}{4} = \frac{13}{6} - \frac{5 \cdot 1}{2 \cdot 4} = \frac{13}{6} - \frac{5}{8} =$$

$$\frac{(24 \div 6) \cdot 13 - (24 \div 8) \cdot 5}{24} =$$

$$\frac{4 \cdot 13 - 3 \cdot 5}{24} = \frac{52 - 15}{24} = \frac{37}{24}$$

Example 6:

$$\frac{11}{4} + 2^2 = \frac{11}{4} + 4 = \frac{(4 \div 4) \cdot 11 + (4 \div 1) \cdot 4}{4} =$$
$$\frac{11 + 16}{4} = \frac{27}{4}$$

Example 7:

$$-\frac{16}{9} \cdot \frac{1}{2} - \left(-2 + \frac{49}{10}\right) = -\frac{16}{9} \cdot \frac{1}{2} - \left(\frac{-10 \cdot 2 + 1 \cdot 49}{10}\right) =$$

$$-\frac{16}{18} - \left(\frac{-20 + 49}{10}\right) = -\frac{16}{18} - \left(\frac{29}{10}\right) =$$

$$-\frac{8}{9} - \frac{29}{10} = \frac{-80 - 261}{90} = -\frac{341}{90}$$

Example 8:

$$\frac{9}{5}\left(\frac{8}{3} - \frac{4}{3}\right) = \frac{9}{5} \cdot \frac{4}{3} = \frac{12}{5}$$

Example 9:

$$-\frac{5}{9} \cdot \frac{4}{5} + \frac{2}{3} \cdot \frac{15}{8} = -\frac{4}{9} + \frac{10}{8} =$$

$$\frac{-32 + 90}{72} = \frac{58}{72} = \frac{29}{36}$$

Problems Solved

1) **Problem 7 :**

$$\frac{13}{6} - \frac{5}{2} \cdot \frac{1}{4}$$

first we do the multiplication and then subtraction, so

$$\frac{13}{6} - \frac{5}{2} \cdot \frac{1}{4} = \frac{13}{6} - \frac{5}{8} = \frac{13(24 \div 6) - 5(24 \div 8)}{24} =$$

$$\frac{13 \cdot 4 - 5 \cdot 3}{24} = \frac{52 - 15}{24} = \mathbf{\frac{37}{24}}$$

2) **Problem 14:**

$$-6 - -6 \cdot 5 - 10$$

first we do the multiplication and then subtraction, so

$$-6 - -6 \cdot 5 - 10 = -6 - -30 - 10 =$$

$$-6 + 30 - 10 = \mathbf{14}$$

3) **Problem 10:**

$$\frac{11}{4} + 2^2$$

first we do the power and then addition, so

$$\frac{11}{4} + 2^2 = \frac{11}{4} + 4 = \frac{11(4 \div 4) + 4(4 \div 1)}{4} =$$

$$\frac{11 \cdot 1 + 4 \cdot 4}{4} = \frac{11 + 16}{4} = \mathbf{\frac{27}{4}}$$

4) **Problem 18 :**

$$-\frac{16}{9}\cdot\frac{1}{2}-\left(-2+\frac{49}{10}\right)$$

$$-\frac{16}{9}\cdot\frac{1}{2}-\left(-2+\frac{49}{10}\right)=$$

first we solve the operations inside the parenthesis, so:

$$-\frac{16}{9}\cdot\frac{1}{2}-\left(\frac{-20+49}{10}\right)=-\frac{16}{9}\cdot\frac{1}{2}-\left(\frac{29}{10}\right)=$$

now we eliminate the parenthesis, so:

$$-\frac{16}{9}\cdot\frac{1}{2}-\frac{29}{10}=$$

now we do the multiplication and then subtraction, so:

$$-\frac{16\div2}{9}\cdot\frac{1}{2\div2}-\frac{29}{10}=-\frac{8}{9}-\frac{29}{10}=$$

$$\frac{-8\cdot(90\div9)-29(90\div10)}{90}=\frac{-8\cdot10-29\cdot9}{90}=$$

$$\frac{-80-261}{90}=-\frac{341}{90}$$

5) Problem 16 :

$$\frac{7}{8} \cdot \frac{1}{9} \cdot \left(\frac{5}{4}\right)^2$$

first we do the operation inside parenthesis,
that is the power, then multiplication, so:

$$\frac{7}{8} \cdot \frac{1}{9} \cdot \frac{25}{16} = \frac{175}{1152}$$

6) Problem 20 :

$$-\frac{5}{9} \cdot \frac{4}{5} + \frac{2}{3} \cdot \frac{15}{8}$$

first we do the multiplication, but if you look well,
you see that we can simplify by criss cross, so:

$$-\frac{5 \div 5}{9} \cdot \frac{4}{5 \div 5} + \frac{2 \div 2}{3 \div 3} \cdot \frac{15 \div 3}{8 \div 2} =$$

$$-\frac{1}{9} \cdot 4 + \frac{1}{1} \cdot \frac{5}{4} = -\frac{4}{9} + \frac{5}{4} =$$

$$\frac{-4(36 \div 9) + 5(36 \div 4)}{36} = \frac{-4 \cdot 4 + 5 \cdot 9}{36} =$$

$$\frac{-16 + 45}{36} = \frac{29}{36}$$

Problems

1	$\dfrac{3+1}{4}$	**11**	$-4 - 8 \cdot -8 - 9$
2	$\left(\dfrac{12^2}{4}\right)$	**12**	$\dfrac{3 \cdot 2}{-3}$
3	$3(5-4)$	**13**	$\dfrac{12}{2} - (8+9)$
4	$6 - 3 - 1$	**14**	$-6 - -6 \cdot 5 - 10$
5	$4 - 1^2$	**15**	$\dfrac{-9-9}{-6+3}$
6	$\dfrac{3}{4} + 1 + 2$	**16**	$\dfrac{7}{8} \cdot \dfrac{1}{9} \cdot \left(\dfrac{5}{4}\right)^2$
7	$\dfrac{13}{6} - \dfrac{5}{2} \cdot \dfrac{1}{4}$	**17**	$10 \cdot \dfrac{7}{2} - \left(\dfrac{43}{8} - -\dfrac{16}{5}\right)$
8	$\dfrac{9}{5}\left(\dfrac{8}{3} - \dfrac{4}{3}\right)$	**18**	$-\dfrac{16}{9} \cdot \dfrac{1}{2} - \left(-2 + \dfrac{49}{10}\right)$
9	$\dfrac{9}{4} - \dfrac{11}{6} + 2$	**19**	$\dfrac{9}{8} - \dfrac{15}{8} - \dfrac{11}{3} - -3$
10	$\dfrac{11}{4} + 2^2$	**20**	$-\dfrac{5}{9} \cdot \dfrac{4}{5} + \dfrac{2}{3} \cdot \dfrac{15}{8}$

EVALUATION EXPRESSIONS

Next, we will go over evaluating expressions. To start with, a variable is a letter which takes the place of an unknown number.

How To Solve

The most used method to solve expression is to substitute numbers in place of the variables.

Example 1: Find the result for:

$$x2 = 6, \quad where \ x = 3$$
$$3(2) = 6$$

Example 2: Find the result for:

$$2 + x2 \quad where \ x = 3$$
$$2 + 3(2) = 8$$

Example 3: Find the result for:

$$y2 + 7 \quad where \ y = 8$$
$$8(2) + 7 = 23$$

Example 4: Find the value of z in the expression :

$$z = 2x(3y), \quad x = 2 \ and \ y = 3$$
$$z = 2 \cdot 2 \cdot 3 \cdot 3 = 36$$

Problems Solved

1) **Problem 11**

$$(c + b)^2 \quad where \; b = \frac{5}{4} \quad c = \frac{11}{3}$$

$$\left(\frac{11}{3} + \frac{5}{4}\right)^2 = \left(\frac{44 + 15}{12}\right)^2 = \left(\frac{59}{12}\right)^2 = \frac{3481}{144}$$

2) **Problem 20**

$$\frac{h^2}{j} \quad where \; h = \frac{11}{3} \quad j = \frac{7}{6}$$

$$\frac{\left(\frac{11}{3}\right)^2}{\frac{7}{6}} = \frac{\frac{121}{9}}{\frac{7}{6}} = \frac{121}{9} \cdot \frac{6}{7} =$$

criss cross simplify since 6 and 9 are divisible by 3, so

$$\frac{121 \cdot 2}{3 \cdot 7} = \frac{242}{21}$$

3) **Problem 18**

$$2 - (k - j) \quad where \; h = \frac{2}{3} \quad j = \frac{7}{6}$$

$$2 - \left(\frac{7}{6} - \frac{2}{3}\right) = 2 - \left(\frac{7 - 4}{6}\right) = 2 - \frac{3}{6} = \frac{12 - 3}{6} =$$

$$\frac{4}{6} = \frac{2}{3}$$

4) Problem 8

$$(x + z)^2 \quad use \ x = 2, \ z = 4$$

$$(2 + 4)^2 \ = \ 6^2 = \mathbf{36}$$

5) Problem 2

$$\frac{a}{6} + c \quad use \ a = 6, \quad c = 2$$

$$\frac{6}{6} + 2 = 1 + 2 = \mathbf{3}$$

6) Problem 10

$$\frac{x + z}{6} \quad use \ x = 4, \ z = 2$$

$$\frac{4 + 2}{6} = \frac{6}{6} = \mathbf{1}$$

7) Problem 9

$$z(x + z) \quad use \ x = 1, \quad z = 2$$

$$2(1 + 2) \ = 2 \cdot 3 = \mathbf{6}$$

8) Problem 6

$$m + \frac{q}{4} \quad use \ m = 5 \quad q = 4$$

$$5 + \frac{4}{4} = 5 + 1 = \mathbf{6}$$

Problems

1	$z + z + y;$ *use* $y = 1,$ $z = 3$
2	$\dfrac{a}{6} + c;$ *use* $a = 6,$ $c = 2$
3	$x + x + y;$ *use* $x = 3,$ $y = 2$
4	$yz - 2;$ *use* $y = 2,$ $z = 6$
5	$q + r - p;$ *use* $p = 3, q = 3, r = 2$
6	$m + \dfrac{q}{4};$ *use* $m = 5,$ $q = 4$
7	$5 - k + j;$ *use* $j = 6,$ $k = 4$
8	$(x + z)^{2};$ *use* $x = 2,$ $z = 4$
9	$z(x + z);$ *use* $x = 1,$ $z = 2$
10	$\dfrac{x + z}{6};$ *use* $x = 4,$ $z = 2$

11	$(c + b)^2$; use $b = \frac{5}{4}$, $c = \frac{11}{3}$
12	$4 + n + p$; use $n = \frac{19}{4}$, $p = \frac{7}{2}$
13	$\frac{q + 6}{r}$; use $q = \frac{7}{6}$, $r = \frac{4}{5}$
14	$6 \cdot \frac{z}{y}$; use $y = \frac{8}{5}$, $z = 1$
15	$y - \frac{z}{4}$; use $y = \frac{7}{4}$, $z = \frac{7}{4}$
16	$a - a + c$; use $a = \frac{11}{6}$, $c = 1$
17	$4 - y + z$; use $y = 2$, $z = \frac{1}{6}$
18	$2 - (k - j)$; use $j = \frac{2}{3}$, $k = \frac{7}{6}$
19	$1 + ba$; use $a = 2$, $b = \frac{1}{2}$
20	$\frac{h^2}{j}$; use $h = \frac{11}{3}$, $j = \frac{7}{6}$

POLYNOMIALS

Multiplying Polynomials

Here we should go into how to multiply polynomials and binomials, as we only went over how to add and subtract them earlier. The first step in multiplying polynomials is to distribute the terms of the first polynomial to those of the second. We should also take to care to combine like terms if there are any.

How To Solve

First, we have to try to distribute the monomial to all of the terms of the polynomial.

Example 1

$$3x^2(4x^2 - 5x + 7) = 12x^4 - 15x^3 + 21x^2$$

Here we would combine like terms if there were any. As you can see, there are none, so our answer here remains

$$12x^4 - 15x^3 + 21x^2$$

Next, we will multiply $-6xy\,(4x^2 - 5xy - 2y^2)$. Here again, we have to try distributing the first term throughout our polynomial following it. This would look like this:

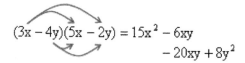

$$-6xy(4x^2 - 5xy - 2y^2) = -24x^3y + 30x^2y^2 + 12xy^3$$

Again, there are no like terms here, so our final answer is:

$$-24x^3y + 30x^2y^2 + 12xy^3$$

Example 2

Here we have two binomials that we need to multiply:

$$(3x - 4y)(5x - 2y)$$

In this case, we would distribute both terms within our first binomial to both terms in the second binomial. This would look like this:

$$(3x - 4y)(5x - 2y) = 15x^2 - 6xy \\ - 20xy + 8y^2$$

We do have some like terms in this answer this time (-6xy and -20xy). After combining these terms, our final answer is:

$$15x^2 - 26xy + 8y^2$$

Problems Solved

1) **Problem 11**

$$(5p - 8)(6p + 7)$$

$$(5p - 8)(6p + 7) = 30p^2 + 35p - 48p - 56 =$$

$$30p^2 + (+35 - 48)p - 56 = \mathbf{30p^2 - 13p - 56}$$

2) **Problem 16**

$$\left(\frac{13}{3}x - \frac{6}{7}\right)\left(\frac{19}{6}x - 1\right)$$

$$\left(\frac{13}{3}x - \frac{6}{7}\right)\left(\frac{19}{6}x - 1\right) = \frac{247}{18}x^2 - \frac{13}{3}x - \frac{114}{42}x + \frac{6}{7} =$$

$$\frac{247}{18}x^2 + \left(-\frac{13}{3} - \frac{114}{42}\right)x + \frac{6}{7} =$$

$$\frac{247}{18}x^2 + \left(\frac{-182 - 114}{42}\right)x + \frac{6}{7} =$$

$$\frac{247}{18}x^2 - \frac{296}{42}x + \frac{6}{7} = \frac{296}{42} =$$

$$\frac{247}{18}x^2 - \frac{148}{21}x + \frac{6}{7}$$

3) **Problem 7**

$$\frac{1}{2}\left(2v + \frac{1}{2}\right)$$

$$\frac{1}{2}\left(2v + \frac{1}{2}\right) = \frac{2}{2}v + \frac{1}{4} = v + \frac{1}{4}$$

4) **Problem 17**

$$\left(2x - \frac{13}{4}\right)\left(2x + \frac{27}{8}\right)$$

$$\left(2x - \frac{13}{4}\right)\left(2x + \frac{27}{8}\right) = 4x^2 + \frac{54}{8}x - \frac{26}{4}x - \frac{351}{32} =$$

$$4x^2 + \left(\frac{54}{8} - \frac{26}{4}\right)x - \frac{351}{32} = 4x^2 + \left(\frac{54 - 52}{8}\right)x - \frac{351}{32} =$$

$$4x^2 + \frac{2}{8}x - \frac{351}{32} = 4x^2 + \frac{1}{4}x - \frac{351}{32}$$

5) **Problem 10**

$$\frac{5}{7}\left(\frac{9}{4}p + \frac{23}{8}\right)$$

$$\frac{5}{7}\left(\frac{9}{4}p + \frac{23}{8}\right) = \frac{45}{28}p + \frac{115}{56}$$

Problems

1	$6(6a - 3)$	**11**	$(5p - 8)(6p + 7)$
2	$4(6r + 7)$	**12**	$(4b + 1)(4b + 7)$
3	$7x(4x + 8)$	**13**	$(3n + 7)(5n + 1)$
4	$7(6p + 7)$	**14**	$(m - 8)(m - 4)$
5	$8n(3n + 7)$	**15**	$(8r - 7)(4r - 3)$
6	$\dfrac{33p}{8}\left(\dfrac{6}{5}p - \dfrac{14}{5}\right)$	**16**	$\left(\dfrac{13}{3}x - \dfrac{6}{7}\right)\left(\dfrac{19}{6}x - 1\right)$
7	$\dfrac{1}{2}\left(2v + \dfrac{1}{2}\right)$	**17**	$\left(2x - \dfrac{13}{4}\right)\left(2x + \dfrac{27}{8}\right)$
8	$\dfrac{1}{2}\left(\dfrac{7}{6}v + 1\right)$	**18**	$\left(\dfrac{17}{8}n - 2\right)\left(\dfrac{9}{5}n - \dfrac{3}{2}\right)$
9	$\dfrac{17}{6}\left(\dfrac{2}{3}n + \dfrac{3}{7}\right)$	**19**	$\left(\dfrac{17}{4}k + \dfrac{4}{3}\right)\left(\dfrac{8}{5}k - \dfrac{9}{5}\right)$
10	$\dfrac{5}{7}\left(\dfrac{9}{4}p + \dfrac{23}{8}\right)$	**20**	$\left(\dfrac{1}{6}r + 1\right)\left(2r + \dfrac{11}{4}\right)$

Adding and Subtracting

For instance, the terms 42, 5x, $14x^{12}$, and 2pq would all be considered monomials, whereas the terms 4+y, 5y, 14x, and 2pq−2 would not because none of these terms meet all the criteria of monomials. The next term we need to define is the degree of a monomial.

<u>The monomials degree is the grand sum of all its variables included. Constant terms have monomial degrees of 0.</u>

Some examples of monomials and their degrees are listed below:

Monomial	Degree
42	0
$5x$	$0 + 1 = 1$
$14x^{12}$	$0 + 12 = 12$
$2pq$	$0 + 1 + 1 = 2$

A **polynomial** is the grand sum of all monomials when all the monomials are called terms.

When we use the expression "**degree of a polynomial**", we are referring to our greatest degree of all of the polynomial's terms.

When written out, polynomials are often arranged by the value of each and every monomial's exponents, with the highest exponents being included first to the left and moving down in value to the right.

Here polynomials first term goes by the name of its leading coefficient.

Here is an example of a polynomial being written out by the order of its exponents:

$$4x^5 + x^2 - 14x + 12$$

A polynomial, as you can see, is nothing more than a fancy word for the sum of multiple monomials. Once we have found the order in which to express our polynomials, we can then add other polynomials to it.

How to Solve

We will now add two polynomials together to get an idea of how this process unfolds:

$$(4x^2 + 3x - 14) + (x^3 - x^2 + 7x + 1)$$

The first step that we need to perform here is to group the polynomials together by their exponents. This would look like this:

$$x^3 + (4x^2 - x^2) + (3x + 7x) + (-14 + 1) =$$
$$x^3 + (4x^2 - x^2) + (3x + 7x) + (-13) =$$
$$x^3 + 3x^2 + 10x + (-13) =$$
$$x^3 + 3x^2 + 10x - 13$$

The same procedure should be applied when subtracting polynomials:

$$(4x^2 + 3x - 14) - (x^3 - x^2 + 7x + 1) =$$
$$4x^2 + 3x - 14 - x^3 + x^2 - 7x - 1 =$$
$$5x^2 + 3x - 14 - x^3 - 7x - 1 =$$
$$5x^2 - 4x - 15 - x^3 =$$
$$-x^3 + 5x^2 - 4x - 15$$

2) Using Add in columns:

$$-4x^3 + x^2y - xy^2$$
$$3x^3 \quad + 5x^2y - xy^2$$
$$7xy^2 + 3y^3$$

$$\overline{-x^3 + 6x^2y + 5xy^2 + 3y^3}$$

Problems Solved

1) **Problem 1**

$$(8n^2 + 5n^3) - (4n^2 - 4n) =$$

$$(8n^2 + 5n^3) - (4n^2 - 4n) = 8n^2 + 5n^3 - 4n^2 + 4n =$$

$$\mathbf{5n^3 + 4n^2 + 4n}$$

2) **Problem 11**

$$\left(\frac{5}{3}n - \frac{19}{7}n^3\right) - \left(\frac{4}{5}n^3 - \frac{8}{5}n\right) =$$

$$\frac{5}{3}n - \frac{19}{7}n^3 - \frac{4}{5}n^3 + \frac{8}{5}n =$$

$$\left(\frac{5}{3} + \frac{8}{5}\right)n + \left(-\frac{19}{7} - \frac{4}{5}\right)n^3 =$$

$$\left(\frac{25 + 24}{15}\right)n + \left(-\frac{95 - 28}{35}\right)n^3 = \frac{49}{15}n - \frac{123}{35}n^3$$

now in order from higher degree to low

$$-\frac{123}{35}n^3 + \frac{49}{15}n$$

3) **Problem 9**

$$(2x^2 + 5x^3) - (3x^3 + 5x^2) =$$

$$2x^2 + 5x^3 - 3x^3 - 5x^2 = \mathbf{2x^3 - 3x^2}$$

4) Problem 11

$$\left(\frac{5}{3}n - \frac{19}{7}n^3\right) - \left(\frac{4}{5}n^3 - \frac{8}{5}n\right) =$$

$$\left(\frac{5}{3}n - \frac{19}{7}n^3\right) - \left(\frac{4}{5}n^3 - \frac{8}{5}n\right) = \frac{5}{3}n - \frac{19}{7}n^3 - \frac{4}{5}n^3 + \frac{8}{5}n =$$

$$\left(-\frac{19}{7} - \frac{4}{5}\right)n^3 + \left(\frac{5}{3} + \frac{8}{5}\right)n =$$

$$\left(\frac{-95 - 29}{35}\right)n^3 + \left(\frac{25 + 24}{15}\right)n =$$

$$-\frac{124}{35}n^3 + \frac{49}{15}n$$

Problems

1	$(8n^2 + 5n^3) - (4n^2 - 4n)$
2	$(3n^4 + 7n^2) + (2n^4 + n^2)$
3	$(8n - 6n^2) - (2n - 5n^2)$
4	$(2n^2 + 2n^4) + (6n^4 + 5n^2)$
5	$(2n^2 + 2n^4) + (6n^4 + 5n^2)$
6	$(1 + 3r^4) - (3r^4 - 2)$
7	$(6v^4 - 7) + (4 + 2v^4)$
8	$(8b^2 - 6b) + (6b + 8b^2)$
9	$(2x^2 + 5x^3) - (3x^3 + 5x^2)$
10	$(3 + 8v^2) - (4 + 2v^2)$
11	$\left(\frac{5}{3}n - \frac{19}{7}n^3\right) - \left(\frac{4}{5}n^3 - \frac{8}{5}n\right)$
12	$\left(k^4 + \frac{7}{4}\right) + \left(\frac{7}{5}k^4 + 1\right)$
13	$\left(\frac{1}{7}b^2 + \frac{4}{5}b^4\right) + \left(\frac{3}{2}b^2 + \frac{2}{7}b^4\right)$

14	$\left(\dfrac{29}{6}r^3 - 2r^4\right) - \left(\dfrac{7}{5}r^4 - 8r^3\right)$
15	$\left(\dfrac{3}{2}b^3 - \dfrac{27}{4}b\right) + \left(\dfrac{1}{3}b^3 + \dfrac{2}{5}b\right)$
16	$\left(\dfrac{13}{7}m^3 + \dfrac{9}{5}m^4\right) + \left(\dfrac{4}{5}m^3 + \dfrac{15}{4}\right)$
17	$\left(\dfrac{12}{5}a + \dfrac{32}{7}a^2\right) + \left(\dfrac{4}{3}a - \dfrac{23}{7}a^2\right)$
18	$\left(\dfrac{19}{8}k^4 + \dfrac{3}{4}k^3\right) - \left(\dfrac{3}{5}k^4 + \dfrac{8}{5}k^3\right)$
19	$\left(\dfrac{1}{2}n - \dfrac{7}{2}n^3\right) - \left(n - \dfrac{10}{3}n^3\right)$
20	$\left(\dfrac{16}{7}n^3 - \dfrac{5}{3}n^4\right) - \left(\dfrac{23}{7}n^4 - \dfrac{3}{5}n^3\right)$

Dividing

Polynomials can sometimes be divided by using the standard methods that are usually applied to dividing fractions without exponents. There are other times, however, that it becomes advisable to use long division on polynomials instead. The top polynomial in these equations is always referred to as the numerator, while the bottom polynomial is always referred to as the denominator. One great way of remembering the difference between the two is by thinking of the denominator as the "down-ominator".

The Remainder Theorem Method

This theorem is helpful in evaluating and solving for polynomials that already have an unknown value of x. Whenever we encounter this theorem, it will always include a polynomial expressed as $P(x)$, in which $P(x)$ just represents some polynomial with a variable expressed as x.

The next step in solving for this theorem involves dividing the polynomial by a linear factor known as x-a, in which a just represents another random variable. The next step for us here would be to perform long polynomial division on the function. This would leave us with the answer q(x), along with some reminder r(x).

Example, we will now divide the polynomial

$$P(x) = x^3 - 7x - 6$$

by our factor

$$x - 4 \text{ (so } a = 4)$$

How to Solve

Polynomials can sometimes be divided by using the standard methods that are usually applied to dividing fractions without exponents. There are other times, however, that it becomes advisable to use long division on polynomials instead. The top polynomial in these equations is always referred to as the numerator, while the bottom polynomial is always referred to as the denominator. One great way of remembering the difference between the two is by thinking of the denominator as the "down-ominator".

Next, we have to put our equation into the form of something being divided by something else. This requires putting the denominator to the left of a line indicating that the numerator is being divided by it:

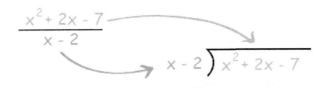

Above is what this process would look like in the case of dividing $x^2 + 2x - 7$ by x - 2.

Both polynomials in these cases (and in all cases for that matter) should feature the highest terms first to the left, and then work their way down the value of their exponents as they move on to the right.

The next steps would be to divide the numerator by the first term featured in the denominator and to put the result of this operation in the answer written above the line placed over the numerator. After this step, we would then multiply our denominator by the new answer found and put the product of this operation below the numerator of the original expression. Finally, we would subtract these two polynomials in order to arrive at our new polynomial.

For example, we will now divide:

$$\frac{x^3 - 4x^2 + 2x - 3}{x + 2}$$

Our first step here would be to check to make sure that these polynomials are written in descending order, which they are in this case, so we do not need to rewrite these polynomials here. Our next step here would be to divide whatever term that has our higher power outside of our division symbol from whatever term that has our higher power inside of our division symbol. Here, we would divide x^3 by the variable x, which would, in turn, leave us with the term x^2, which we would then place on top of the line indicating the division:

$$x + 2 \overline{)x^3 - 4x^2 + 2x - 3}^{\quad x^2}$$

Our next step would be to multiply (or possibly to distribute) the answer that we have arrived at in our previous step by our polynomial placed before our division symbol. In this case, this operation would be multiplying x^2 by $x + 2$. After this operation has been performed, we would place its product $(x^3 + 2x^2)$ below the first two terms within the division symbol.

Our next step would then be to subtract our former from our latter and then to carry down our next term:

$$
\begin{array}{r}
x^2 \\
x + 2 \overline{)x^3 - 4x^2 + 2x - 3} \\
\underline{x^3 + 2x^2 } \\
-6x^2 + 2x
\end{array}
$$

Our next step would be to now divide that term which has our higher power inside of our symbol for division by that term which has our higher power outside of our symbol for division. In the present case, this step would consist of dividing $-6x^2 + 2x$ by x, which would here happen to be -6x:

$$\begin{array}{r} x^2 - 6x \\ x+2\overline{\smash{\big)}\,x^3 - 4x^2 + 2x - 3} \\ \underline{x^3 + 2x^2} \\ -6x^2 + 2x \end{array}$$

Our next step would then be to multiply (or, again, distribute) the answer that we have arrived at in our first step by our polynomial placed in the front of our symbol of division. In the present case, this step would consist of multiplying $-6x \cdot (x + 2)$:

$$\begin{array}{r} x^2 - 6x \\ x+2\overline{\smash{\big)}\,x^3 - 4x^2 + 2x - 3} \\ \underline{x^3 + 2x^2} \\ -6x^2 + 2x \\ -6x^2 - 12x \end{array}$$

Here again, in this next step, we would have to subtract as well as bring down our next term:

$$\begin{array}{r} x^2 - 6x \\ x+2\overline{)x^3 - 4x^2 + 2x - 3} \\ x^3 + 2x^2 \\ \hline -6x^2 + 2x \\ -6x^2 - 12x \\ \hline 14x - 3 \end{array}$$

Our next step would consist of dividing the term which has our higher power inside of our symbol for division by the term which has our higher power outside of our symbol for division. In which case, we would have to divide $14x \div x = 14$.

$$\begin{array}{r} x^2 - 6x + 14 \\ x+2\overline{)x^3 - 4x^2 + 2x - 3} \\ x^3 + 2x^2 \\ \hline -6x^2 + 2x \\ -6x^2 - 12x \\ \hline 14x - 3 \end{array}$$

In our next step, we would need to distribute or multiply our answer which was found in our first step by our polynomial placed before our symbol for division. In the present case, we would need here to times 14 *by x* + 2:

$$
\begin{array}{r}
x^2 - 6x + 14 \\
x + 2 \overline{\smash{\big)}\, x^3 - 4x^2 + 2x - 3} \\
\underline{x^3 + 2x^2\phantom{{}+2x-3}} \\
-6x^2 + 2x\phantom{{}-3} \\
\underline{-6x^2 - 12x\phantom{{}-3}} \\
14x - 3 \\
14x + 28
\end{array}
$$

Our next and penultimate step would be to subtract again. Upon doing so, we notice that there are no more terms to bring down:

$$
\begin{array}{r}
x^2 - 6x + 14 \\
x + 2 \overline{\smash{\big)}\, x^3 - 4x^2 + 2x - 3} \\
\underline{x^3 + 2x^2\phantom{{}+2x-3}} \\
-6x^2 + 2x\phantom{{}-3} \\
\underline{-6x^2 - 12x\phantom{{}-3}} \\
14x - 3 \\
14x + 28 \\
\underline{-31}
\end{array}
$$

Now we would just need to write our final answer. The term that the final subtraction operation leaves us with is known as our remainder which should be expressed as one fraction in our last answer:

$$
x^2 - 6x + 14 - \frac{31}{x + 2}
$$

Problems Solved

1) Problem 1

$$\frac{v^2 + 4v - 45}{v - 5}$$

to solve you can apply the long division by polynomial,
as explained in the previous pages, or
if you look well, between v^2 and v, we can put v
as commom factor, the number 45 can be
factorized as $5 \cdot 9$, so:

$$\frac{v^2 + 4v - 45}{v - 5} = \frac{v(v + 4) - 5 \cdot 9}{v - 5} = \frac{(v - 5)(v + 9)}{(v - 5)} = v + 9$$

2) Problem 16

$$\frac{x^2 - 6x - 7}{x - 7}$$

if you look well, between x^2 and $-6x$, we can put x
as commom factor, and the number 7 can be
factorized as $7 \cdot 1$, so:

$$\frac{x(x - 6) - 7 \cdot 1}{x - 7} =$$

$$\frac{(x - 7) \cdot (x + 1)}{x - 7} = x + 1$$

3) Problem 18

*if you look well, between x^2 and $+8x$, we can put x
as commom factor, and the number 20 can be
factorized as $10 \cdot 2$, so:*

$$\frac{x^2 + 8x - 20}{x - 2} = \frac{x(x + 8) + 10 \cdot -2}{x - 2} =$$

$$\frac{(x + 10) \cdot \cancel{(x - 2)}}{\cancel{x - 2}} = x + 10$$

4) Problem 19

*if you look well, between x^2 and $-9x$, we can put x
as commom factor, and the number 14 can be
factorized as $7 \cdot 2$, so:*

$$\frac{x^2 - 9x + 14}{x - 7} = \frac{x(x - 9) - 2 \cdot -7}{x - 7} =$$

$$\frac{(x - 2) \cdot \cancel{(x - 7)}}{\cancel{x - 7}} = x - 2|$$

Problems

1	$\dfrac{v^2 + 4v - 45}{v - 5}$	**11**	$\dfrac{5n^2 - 47n - 30}{n - 10}$
2	$\dfrac{r^2 - 3r - 54}{r + 6}$	**12**	$\dfrac{r^2 - 19r + 90}{r - 9}$
3	$\dfrac{a^2 + 5a - 6}{a + 6}$	**13**	$\dfrac{n^2 - 12n + 35}{n - 7}$
4	$\dfrac{b^2 + b - 6}{b + 3}$	**14**	$\dfrac{m^2 - 16m + 63}{m - 7}$
5	$\dfrac{k^2 - 7k + 12}{k - 4}$	**15**	$\dfrac{v^2 + 4v + 4}{v + 2}$
6	$\dfrac{m^2 + 8m + 12}{m + 6}$	**16**	$\dfrac{x^2 - 6x - 7}{x - 7}$
7	$\dfrac{n^2 - 6n + 5}{n - 5}$	**17**	$\dfrac{n^2 - 5n + 4}{n - 4}$
8	$\dfrac{4k^2 + 14k + 12}{k + 2}$	**18**	$\dfrac{x^2 + 8x - 20}{x - 2}$
9	$\dfrac{4n^2 - 18n + 8}{n - 4}$	**19**	$\dfrac{x^2 - 9x + 14}{x - 7}$
10	$\dfrac{9x^2 + 15x - 6}{x + 2}$		

BINOMIALS

Square of Binomials

$$(a + b)^2 = a^2 + 2ab + b^2$$

$$(a - b)^2 = a^2 - 2ab + b^2$$

How to solve

Example 1: case $(a + b)^2 = a^2 + 2ab + b^2$

$$(2x + 6)^2 = (2x)^2 + 2(2x)(6) + (6)^2 = 4x^2 + 24x + 36$$

Example 2: case $(a - b)^2 = a^2 - 2ab + b^2$

$$(1 - 7x)^2 = (1)^2 - 2(1)(7x) + (7x)^2 = 1 - 14x + 49x^2$$

Example 3: case $(a + b)^2 = a^2 + 2ab + b^2$

$$\left(\frac{1}{4}x^2 + \frac{29}{8}\right)^2 = \left(\frac{1}{16}x^4 + 2 \cdot \frac{1}{4}x^2 \cdot \frac{29}{8} + \frac{841}{64}\right) =$$

$$\left(\frac{1}{16}x^4 + \frac{58}{32}x^2 + \frac{841}{64}\right) =$$

if you look well, we can simplify the fraction $\frac{58}{32}$ dividing by 2

$$\frac{1}{16}x^4 + \frac{(58 \div 2)}{(32 \div 2)}x^2 + \frac{841}{64} = \frac{1}{16}x^4 + \frac{29}{16}x^2 + \frac{841}{64}$$

Problems Solved

1) **Problem 15**

$$(m + 3)^2$$

if you look well, we can apply the formula

$$(a + b)^2 = a^2 + 2ab + b^2$$

in this case: $a = m, \ b = 3$

$$(m + 3)^2 = m^2 + 2(m)(3) + 9 = m^2 + 6m + 9$$

2) **Problem 16**

$$\left(2x + \frac{1}{6}\right)^2$$

if you look well, we can apply the formula

$$(a + b)^2 = a^2 + 2ab + b^2$$

in this case: $a = 2x, \ b = \frac{1}{6}$

$$\left(2x + \frac{1}{6}\right)^2 = 4x^2 + 2\left(2x \cdot \frac{1}{6}\right) + \left(\frac{1}{6}\right)^2 =$$

$$4x^2 + \frac{4}{6}x + \frac{1}{36} =$$

$$4x^2 + \frac{2}{3}x + \frac{1}{36}$$

3) Problem 7

$$\left(x - \frac{1}{2}\right)^2$$

if you look well, we can apply the formula

$$(a - b)^2 = a^2 - 2ab + b^2$$

in this case: $a = x, \ b = -\frac{1}{2}$

$$x^2 - 2(x)\left(\frac{1}{2}\right) + \left(\frac{1}{2}\right)^2 =$$

$$x^2 - x + \frac{1}{4}$$

Problems

1	$(b-2)^2$	**11**	$(3x-7)^2$
2	$(n-4)^2$	**12**	$(2x+2)^2$
3	$(n-1)(n+1)$	**13**	$(6n-8)^2$
4	$(r+5)^2$	**14**	$(4b+2)^2$
5	$(x-2)(x+2)$	**15**	$(m+3)^2$
6	$\left(x-\dfrac{5}{3}\right)^2$	**16**	$\left(2x+\dfrac{1}{6}\right)^2$
7	$\left(x-\dfrac{1}{2}\right)^2$	**17**	$\left(-3+\dfrac{5}{3}x\right)^2$
8	$\left(k-\dfrac{6}{5}\right)^2$	**18**	$\left(\dfrac{1}{4}x^2+\dfrac{29}{8}\right)^2$
9	$\left(a-\dfrac{7}{5}\right)^2$	**19**	$\left(1+\dfrac{3}{4}p\right)^2$
10	$\left(k-\dfrac{9}{5}\right)^2$	**20**	$\left(\dfrac{3}{4}x^3+\dfrac{1}{2}\right)^2$

Differences of two Squares

Hereafter the following different cases:

$$(a + b)(a - b) = a^2 - b^2$$

$$(a - b)(a + b) = a^2 - b^2$$

$$(a + b)(-a + b) = -a^2 + b^2$$

$$(-a + b)(a + b) = -a^2 + b^2$$

How to solve

Next, we should touch on the difference of two squares pattern for factoring binomials.

The first step in order to start factoring polynomials is always to factor out whatever common factors you happen to find.

The difference of two squares pattern is a special of factoring out binomials that looks like this:

$$x^2 - y^2 = (x + y)(x - y)$$

$$(x + y)(x - y) = x^2 - y^2$$

There are some requirements that need to be met in order to use this pattern though, these are as follows: it must be a binomial that we are dealing with, both products have to be perfect squares, and there must be a subtraction sign in between the two terms. These are the steps to factoring the binomial if all these requirements have been met:

1) Write two parentheses
2) Put a + in one and a - in the other
3) Use the square root for our primary term and put that in the front of each parenthesis.
4) Take the square root for our last term and put that in the back of each parenthesis

Example 1:

$$x^2 - 25$$

First, we would arrange the square roots of both terms into two sets of parentheses that include a subtraction and an addition sign: (x + 5) (x - 5)

Next, we would distribute these two expressions into one another, which would leave us with

$$x^2 - 5x + 5x - 25 = x^2 - 25$$

$$(5x + 3)(5x - 3) = (5x)^2 - (3)^2 = 25x^2 - 9$$

Example 2:

$$(x + y)(x - y)$$

this is the case $(a + b)(a - b) = a^2 - b^2$

$$(x + y)(x - y) = x^2 - y^2$$

Example 3:

$$(2u + 2)(2u - 2)$$

this is the case $(a + b)(a - b) = a^2 - b^2$

$$(2u + 2)(2u - 2) = (2u)^2 - 2^2 = 4u^2 - 4$$

Problems Solved

1) **Problem 11**

$$\left(\frac{1}{3}x + \frac{25}{8}\right)\left(\frac{1}{3}x - \frac{25}{8}\right)$$

if you look well, we can apply the formula

$$(a+b)(a-b) = a^2 - b^2$$

$$a = \frac{1}{3}x, \qquad b = \frac{25}{8}$$

now we do $a^2 - b^2$, so:

$$\left(\frac{1}{3}x\right)^2 - \left(\frac{25}{8}\right)^2 = \frac{1}{9}x^2 - \frac{625}{64}$$

2) **Problem 18**

$$\left(-2 + \frac{22}{5}x\right)\left(2 + \frac{22}{5}x\right)$$

if you look well, we can apply the formula

$$(-a+b)(a+b) = -a^2 + b^2$$

$$a = 2, \qquad b = \frac{22}{5}x$$

now we do $-a^2 + b^2$, so:

$$-(2)^2 + \left(\frac{22}{5}x\right)^2 = -4 + \frac{484}{25}x^2$$

3) Problem 20

$$\left(1 + \frac{9}{8}x\right)\left(-1 + \frac{9}{8}x\right)$$

if you look well, we can apply the formula

$$(a + b)(-a + b) = -a^2 + b^2$$

$$a = 1, \qquad b = \frac{9}{8}x$$

now we do $-a^2 + b^2$*, so:*

$$-(1)^2 + \left(\frac{9}{8}x\right)^2 = -1 + \frac{81}{64}x^2$$

4) Problem 5

$$(7x - 5)(7x + 5)$$

if you look well, we can apply the formula

$$(a - b)(a + b) = a^2 - b^2$$

$$a = 7x, \qquad b = 5$$

now we do $a^2 - b^2$*, so:*

$$(7x)^2 - (5)^2 = 49x^2 - 25$$

Problems

1	$(1 + 8p)(1 - 8p)$	**11**	$\left(\dfrac{1}{3}x + \dfrac{25}{8}\right)\left(\dfrac{1}{3}x - \dfrac{25}{8}\right)$
2	$(6x + 6)(6x - 6)$	**12**	$\left(-1 + \dfrac{8}{5}n\right)\left(1 + \dfrac{8}{5}n\right)$
3	$(7k + 3)(7k - 3)$	**13**	$\left(\dfrac{5}{7} - \dfrac{13}{4}p\right)\left(\dfrac{5}{7} + \dfrac{13}{4}p\right)$
4	$(4x - 1)(4x + 1)$	**14**	$\left(\dfrac{15}{8} + \dfrac{31}{8}n\right)\left(\dfrac{15}{8} - \dfrac{31}{8}n\right)$
5	$(7x - 5)(7x + 5)$	**15**	$\left(\dfrac{1}{4}m + \dfrac{3}{2}\right)\left(\dfrac{1}{4}m - \dfrac{3}{2}\right)$
6	$(7x - 2)(7x + 2)$	**16**	$\left(2v + \dfrac{2}{3}\right)\left(-2v + \dfrac{2}{3}\right)$
7	$(3 + 3b)(3 - 3b)$	**17**	$\left(\dfrac{3}{2}n - \dfrac{7}{5}\right)\left(\dfrac{3}{2}n + \dfrac{7}{5}\right)$
8	$(7x + 5)(7x - 5)$	**18**	$\left(-2 + \dfrac{22}{5}x\right)\left(2 + \dfrac{22}{5}x\right)$
9	$(8x + 2)(8x - 2)$	**19**	$\left(n - \dfrac{11}{5}\right)\left(n + \dfrac{11}{5}\right)$
10	$(2p - 8)(2p + 8)$	**20**	$\left(1 + \dfrac{9}{8}x\right)\left(-1 + \dfrac{9}{8}x\right)$

FACTORING

The Terms indicate sum or difference, Factors indicates product.

An expression is in factored form if the entire expression is expressed as a product

In Factored Form	Not in Factored Form
$2x(x + y)$	$2x + 3y + z$
$(x + y)(3x - 2y)$	$2(x + y) + z$
$(x + 4)(x^2 + 3x - 1)$	$(x + y)(2x - y) + 5$

Multiple of a number

We get a multiple of a number when we multiply it by another number. Such as multiplying by 1, 2, 3, 4, 5, etc, but not zero. Just like the multiplication table.

Here are some examples:

- The multiple of 3 are: 3,6,9,12, 15 ....
- The multiple of 4 are: 4,8,12, 16,20 ....

Factorization and Prime Numbers

To start off with, prime numbers are the numbers which are not divided by any other numbers besides either 1 or themselves.

The first few and most common prime numbers there are two, three, five, seven, eleven, thirteen, and seventeen.

Factors include any numbers that you would multiply together to arrive at different numbers.

The process of prime factorization involves finding out which prime numbers can be multiplied together in order to arrive at the original.

For instance, the prime factors of 180 include 2 x 2 x 5 x 3 x3 because these are the smallest prime numbers that can be multiplied to equal 180.

Dividend	Divisor (Factors)
180	2
90	2
45	5
9	3
3	3
1	

Factoring a perfect square of a Trinomial

To factor a perfect square trinomial form with the square of the binomial follow the schema:

$$25x^2 + 20x + 4$$

The square root of $25x^2 = 5x$

The square root of $4 = 2$

The related binomial is $(5x + 2)^2$

Factoring By grouping

It means we "grouped" two or more terms at a time

$$ax - ay - 2x + 2y = a(x - y) - 2(x - y) = (x - y)(a - 2)$$

How to Solve

1) Factor the following polynomial by grouping

$$7x + 7x^3 + x^4 + x^6$$

First step: group the first two term and the last two terms

$$(7x + 7x^3) + (x^4 + x^6)$$

we can factor a $7xx$ out of the first grouping and x^4 out of the second grouping, so

$$7x + 7x^3 + x^4 + x^6 = 7x(1 + x^2) + x^4(1 + x^2)$$

Now we can factor an $x(1 + x^2)$ so we get

$$x(1 + x^2)(7 + x^3)$$

2) Factor the following polynomial by grouping

$$ax - ay + 2x - 2y$$

$$a(x - y) + 2(x - y) = (x - y)(a + 2)$$

3) Factor the following polynomial by grouping

$$2ax + 3a + 4x + 6$$

$$a(2x + 3) + 2(2x + 3) = (2x + 3)(a + 2)$$

4) Factor the following polynomial by grouping

Remember:

$$ac + ad + bc + bd =$$
$$a(c + d) + b(c + d) =$$
$$(a + b)(c + d)$$

$$18x^3 + 9x^2 + 8x + 4 =$$

$$9x^2(2x + 1) + 4(2x + 1) =$$

$$(9x^2 + 4)(2x + 1)$$

5) Factor the following polynomial by grouping

$$16f^3 - 8f^2 - 14f + 7 =$$

$$8f^2(2f - 1) - 7(2f - 1) =$$

$$(8f^2 - 7)(2f - 1)$$

Problems Solved

1) **Problem 1**

$$2b^3 - 5b^2 + 6b - 15$$

we can put b^2 as common factor between $2b^3$ and $5b^2$,

and 3 as common factor between $6b$ and 15, so:

$$2b^3 - 5b^2 + 6b - 15 = b^2(2b - 5) + 3(2b - 5) =$$

$$(\boldsymbol{b^2 + 3})(\boldsymbol{2b - 5})$$

2) **Problem 6**

$$3x^3 - 12x^2 + 4x - 16$$

we can put $3x^2$ as common factor between

$3x^3$ and $12x^2$ and 4 as common factor between

$4x$ and 16

$$3x^3 - 12x^2 + 4x - 16 = 3x^2(x - 4) + 4(x - 4) =$$

$$(\boldsymbol{3x^2 + 4})(\boldsymbol{x - 4})$$

3) Problem 12

$$10n^3 - 40n^2 + 6n - 24$$

we can put $10n^2$ as common factor between
$10n^3$ and $40n^2$ and 6 as common factor between
$6n$ and 24

$$10n^3 - 40n^2 + 6n - 24 = 10n^2(n - 4) + 6(n - 4) =$$

$$(10n^2 + 6)(n - 4) =$$

if you look well $(10n^2 + 6)$, we can factorize too,
because 2 is common factor between $10n^2$ and 6, so
the final solution is

$$\mathbf{2(5n^2 + 3)(n - 4)}$$

4) Problem 2

$$x^3 - 5x^2 + 4x - 20$$

we can put x^2 as common factor between
x^3 and $5x^2$ and 4 as common factor between
$4x$ and 20

$$x^2(x - 5) + 4(x - 5) = (\mathbf{x^2 + 4)(x - 5)}$$

Problems

1	$25b^3 - 5b^2 + 6b - 15$
2	$x^3 - 5x^2 + 4x - 20$
3	$6x^3 - 15x^2 + 2x - 5$
4	$8p^3 - 10p^2 + 20p - 25$
5	$4x^3 - 10x^2 + 2x - 5$
6	$3x^3 - 12x^2 + 4x - 16$
7	$9r^3 + 12r^2 + 6r + 8$
8	$r^3 - 4r^2 + 5r - 20$
9	$20k^3 + 25k^2 + 8k + 10$
10	$15r^3 + 6r^2 + 10r + 4$

11	$24n^3 - 16n^2 + 18n - 12$
12	$10n^3 - 40n^2 + 6n - 24$
13	$10x^3 - 20x^2 + 6x - 12$
14	$18r^3 + 6r^2 + 24r + 8$
15	$8k^3 - 8k^2 + 12k - 12$
16	$9n^3 + 24n^2 + 6n + 16$
17	$32x^3 + 8x^2 + 20x + 5$
18	$8n^3 - 2n^2 + 4n - 1$
19	$24x^3 + 192x^2 + 15x + 120$
20	$42x^3 - 98x^2 - 30x + 70$

Common Factor only

1) A number is divisible by 2 if the last number is even

2) A number is divisible by 5 if the last number is 0 or 5

3) A number is divisible by 10 if the last numbers is 0

How to Solve

Below there are some of Problems solved and explained among those proposed in the next chapter

1) **Problem 1**

$$18b^{10} + 14b^2 - 10 =$$

The minimum common factor is 2, because each term is divisible by 2, so we divide each term by 2 and multiply all by 2, so:

$$2\left(\frac{18b^{10}}{2} + \frac{14b^2}{2} - \frac{10}{2}\right) = 2(9b^{10} + 7b^2 - 5)$$

2) **Problem 8**

$$-70x^5 + 50x - 80 =$$

The minimum common factor is 10, because each term is divisible by 10, so we divide each term by 10 and multiply all by 10, so

$$10\left(\frac{-70x^5}{10} + \frac{50x}{10} - \frac{80}{10}\right) = 10(-7x^5 + 5x - 8)$$

Problem 2

$$80b^5 - 40b^4 + 64b^2$$

The minimum common factor is $8b^2$, because each term is divisible by $8b^2$, so we divide each term by $8b^2$ and multiply all by $8b^2$

$$8b^2\left(\frac{80b^5}{8b^2} - \frac{40b^4}{8b^2} + \frac{64b^2}{8b^2}\right) = 8b^2(10b^3 - 5b^2 + 8)$$

3) ## Problem 4

$$20v^{11} + 36v^4 + 8v^3$$

The minimum common factor is $4v^3$, because each term is divisible by $4v^3$, so we divide each term by $4v^3$ and multiply all by $4v^3$, so

$$4v^3\left(\frac{20v^{11}}{4v^3} + \frac{36v^4}{4v^3} + \frac{8v^3}{4v^3}\right) = 4v^3(5v^8 + 9v + 2)$$

Problems

1	$18b^{10} + 14b^2 - 10$
2	$80b^5 - 40b^4 + 64b^2$
3	$-56 - 56v + 16v^3$
4	$20v^{11} + 36v^4 + 8v^3$
5	$21x^2 + 56x + 63$
6	$42r^6 + 42r^3 - 48r$
7	$15x^7 - 20x^5 + 15x^4$
8	$-70x^5 + 50x - 80$
9	$-18m^8 + 6m^5 - 30m^3$
10	$10p^4 - 14p^2 - 6p$

11	$45x^5 + 90x^3 + 18x^2$
12	$-50a^5 + 90a^3 + 90a^2$
13	$24n - 36 + 28n^4$
14	$-54 + 36r - 9r^3$
15	$12n^3 + 24n + 30$
16	$-4x + 3x^2 + 5x^4$
17	$12n^5 + 18n^3 - 14n^2$
18	$-28b^3 - 63b^2 - 42b$
19	$-63 + 70p^2 - 56p^4$
20	$-6n^6 - 42n^4 - 6n^3$

Greatest Common Factor GCF

The GCF is the biggest number that every one of the numbers in a set of two or more numbers can be divided by.

For example, 4 would be the GCF between 12 and 16.

This is because 4 is the biggest unit which both 12 and 16 can be divided by.

As you can tell already, both 12 and 16 can also be divided by 2, but this is not the greatest common factor because it is less 4.

The Greatest Common Factor (GCF) is the largest whole number that is a factor of each of two more numbers.

How to Solve

1) Write the prime factorization for each number.

2) Find the greater common factor belonging in both of the numbers, so how many times there are the same factors in both numbers.

Example 1 : Find the GCF of the numbers 3 and 5.
   1) Factor each numer, so:

   $3 = \mathbf{3}$ because 3 is a prime number

   $15 = \mathbf{3} \times 5$

   2) Find the common factor belonging in both of the numbers

   The common factor of 3 and 15 is 3, so the GCF = 3

Example 2 : Find the GCF of the numbers 18 and 3.
   1) Factor each numer, so:

   $18 = \mathbf{3} \times 3 \times 2$ because 3 is a prime number

   $3 = \mathbf{3}$

Example 3: Find the GCF of the numbers 36 and 54.
1) Factor each numer, so:

   Factor of 36: $\mathbf{2} \, x \, 2 \, x \, \mathbf{3} \, x \, \mathbf{3}$

   Factors of 54: $\mathbf{3} \, x \, \mathbf{3} \, x \, 3 \, x \, \mathbf{2}$

2) Identify the common factor belonging in both of the numbers

   In this case the common factors are two times 3 and 1 time 2, so

   GCF = **2 x 3 x 3 = 18**

1) The first step to factoring is to factor out the greatest common factor GCF from each term

$$12y^3 - 15y^2 + 6y$$
$$12y^3 = 2 \cdot 2 \cdot \mathbf{3} \cdot \mathbf{y} \cdot y \cdot y$$
$$15y^2 = 5 \cdot \mathbf{3} \cdot \mathbf{y} \cdot y$$
$$6y = 2 \cdot \mathbf{3} \cdot \mathbf{y}$$
$$\text{GCF} = 3y$$
$$\text{so } 3y(4y^2 - 5y + 2)$$

2) Find the greatest common factor from the following polynomial.

$$6x^7 + 3x^4 - 9x^3$$

In this case, we can factor a 3 and x^3 out of each term and so the greatest common factor is $3x^3$

$$6x^7 + 3x^4 - 9x^3 = 3x^3 (2x^4 + x - 3)$$

Problems

Find GCF for the following exercises:

1) $3x^3 + 27x^2 + 9x$

2) $36x^2 - 64y^4$

3) GCF between 81 and 9

4) GFC of the polynomial:

$$6x^2y + 14xy^2 - 42xy - 2x^2y^2$$

QUADRATIC EXPRESSION

Factoring Quadratic Polynomials

Quadratic is term for second degree polynomial, where the largest exponent in a quadratic polynomial is 2.

$$ax^2 + bx + c$$

How to Solve

Example 1:

$$x^2 + 3x + 2$$

$$where \ a = 1, b = 3 \quad c = 2$$

We have to split the number b (3) in two numbers N1 and N2

So that

$$N1 + N2 = b = 3$$

$$N1 \cdot N2 = ac = 2$$

$$so \quad N1 = 1, \quad N2 = 2$$

$$x^2 + (N1 + N2)x + c =$$

$$x^2 + (1 + 2)x + 2=$$

$$x^2 + x + 2x + 2 =$$

now we have two groups $(x^2 + x),$ *and* $(2x + 2)$

$$x(x + 1) + 2(x + 1)=$$

$$(x + 1)(x + 2)$$

Example 2:

$$x^2 + x - 6$$

where $a = 1, \ b = 1, \ c = -6$ a

$$N1 + N2 = b = 1$$

$$N1 \cdot N2 = ac = -6$$

so $N1 = -2, \ N2 = 3$

$$x^2 + (N1 + N2)x + c =$$

$$x^2 + (-2 + 3)x - 6 =$$

$$x^2 - 2x + 3x - 6 =$$

now we have two groups $(x^2 - x),$ *and* $(3x - 6)$

$$x(x - 2) + 3(x - 2) =$$

$$(x - 2)(x + 3)$$

Problems Solved

1) **Problem 14**

$$2x^2 + 17x + 21$$

$$where \ a = 2, \ b = 17, \ c = 21$$

$$N1 + N2 = b = 17$$

$$N1 \cdot N2 = ac = 42$$

$$so \ N1 = 3, \ N2 = 14$$

$$x^2 + (N1 + N2)x + c=$$

$$x^2 + (3 + 14)x + 21=$$

$$x^2 + 3x + 14x + 21$$

now we have two groups $(x^2 + 3x)$ *and* $(14x + 21)$*, but*

14 and 21 of the second group are divisible by 7, so:

$$x(x + 3) + 7(2x + 3)=$$

Solution : $\ \ (2x + 3)(x + 7)$

2) **Problem 11**

$$7b^2 - 8b=$$
This is very simple, we can group immediately
$$b(7b - 8)$$

3) Problem 16

$$20k^2 + 192k - 80$$

$$where \ a = 20, \ \ b = 192, \ \ c = -80$$

$$N1 + N2 = b = 192$$

$$N1 \cdot N2 = ac = -1600$$

$$so \ \ N1 = 20, \ \ N2 = -80$$

$$k^2 + (N1 + N2)k + c=$$

$$k^2 + (20 - 80)k - 80=$$

$$k^2 + 20k - 80k - 80$$

$$now \ we \ can \ gropus \ (k^2 + 20k), and \ (-80k - 80), so$$

$$k(k + 20) - 80(k + 1)=$$

$$(k - 80)(k + 20) =$$

$$(k - 2 \cdot 4 \cdot 10)(k + 2 \cdot 10) =$$

$$k(-4 \cdot 5 \cdot 4)(k + 2 \cdot 5) =$$

$$\mathbf{4(5k - 2)(k + 10)}$$

Problems

1	$a^2 - 20a + 100$	**11**	$7b^2 - 8b$
2	$b^2 + b - 90$	**12**	$15a^2 - 66a - 45$
3	$p^2 - 2p - 8$	**13**	$7n^2 + 58n + 16$
4	$n^2 - 4n + 3$	**14**	$2x^2 + 17x + 21$
5	$x^2 - 4x - 32$	**15**	$14n^2 - 118n + 140$
6	$n^2 + 14n + 45$	**16**	$20k^2 + 192k - 80$
7	$x^2 - 15x + 54$	**17**	$25k^2 - 105k - 270$
8	$3b^2 - 15b - 150$	**18**	$6a^2 + 28a + 32$
9	$5m^2 + 55m + 120$	**19**	$8x^2 + 52x - 180$
10	$n^2 - 7n - 18$	**20**	$20a^2 - 92a + 48$

RADICAL EXPRESSIONS

Radical expressions are, as their name would suggest, expressions that include radicals.

Hereafter the properties to simplify radical expression:

1) **The square root of a product** equals the product of the square roots of the factors.

$$\sqrt{xy} = \sqrt{x} \cdot \sqrt{y}$$

$$\sqrt{\frac{x}{y}} = \frac{\sqrt{x}}{\sqrt{y}} \quad where \ x \geq 0, \qquad y \geq 0$$

You know that

$$\sqrt{x^2} = x \quad where \ x \geq 0$$

A radical expression is in its simplest form if there are no perfect square factors other than 1 in the radicand

Example 1 :

$$\sqrt{9x} = \sqrt{9} \cdot \sqrt{x} = \sqrt{3^2} \cdot \sqrt{x} = 3\sqrt{x}$$

Example 2 :

$$\sqrt{\frac{49}{16}x^2} = \frac{\sqrt{49}}{\sqrt{16}}\sqrt{x^2} = \frac{5}{4}x$$

Example 3 :

$$\sqrt{\frac{15}{81}} = \frac{\sqrt{15}}{\sqrt{81}} = \frac{\sqrt{15}}{9}$$

Example 4 :

If the denominator is not a perfect square you can multiply the expression in a right form,

$$\sqrt{\frac{15}{7}} = \frac{\sqrt{15}}{\sqrt{7}} \cdot \frac{\sqrt{7}}{\sqrt{7}} = \frac{\sqrt{15 \cdot 7}}{\sqrt{7^2}} = \frac{\sqrt{105}}{7}$$

Product of Conjugates Binomials

$$x\sqrt{y} + z\sqrt{w} \quad and \quad x\sqrt{y} - z\sqrt{w}$$

Are conjugates to each other

The following example shows how to multiply two conjugates binomials:

Example 5 :

$$\frac{x}{3 + \sqrt{x}} = \frac{x\left(3 - \sqrt{x}\right)}{\left(3 + \sqrt{x}\right)\left(3 - \sqrt{x}\right)} = \frac{x\left(3 - \sqrt{x}\right)}{9 - \left(\sqrt{x}\right)^2} = \frac{3x - x\sqrt{x}}{9 - x}$$

Adding and Subtracting

Distributive property:

$$a\sqrt{c} + b\sqrt{c} = (a + b)\sqrt{c}$$

How to Solve

Example 1

$$3\sqrt{2} - 10\sqrt{2} = (3 - 10)\sqrt{2} = -7\sqrt{2}$$

Example 2

$$6\sqrt{3} + 3\sqrt{3} = (6 + 3)\sqrt{3} = 9\sqrt{3}$$

Example 3

$$6\sqrt{12} - 12\sqrt{3} + \sqrt{100}$$

$$6\sqrt{4}\ \sqrt{3} - 12\sqrt{3} + \sqrt{100}$$

$$6 \cdot 2\sqrt{3} - 12\sqrt{3} + 10$$

$$12\sqrt{3} - 12\sqrt{3} + 10 = 10$$

Problems Solved

1) **Problem 1**

$$2\sqrt{2} - \sqrt{18} =$$

$$2\sqrt{2} - \sqrt{3 \cdot 3 \cdot 2} = 2\sqrt{2} - \sqrt{3^2 \cdot 2} =$$

$$2\sqrt{2} - 3\sqrt{2} \cdot =$$

now apply the distributive property

$$(2 - 3)\sqrt{2} = -1\sqrt{2} = -\sqrt{\mathbf{2}}$$

2) **Problem 11**

$$-\sqrt{2} - 3\sqrt{5} - 2\sqrt{5} =$$

now apply the distributive property

$$-\sqrt{2} + (-3 - 2)\sqrt{5} =$$

$$-\sqrt{2} - 5\sqrt{5}$$

3) **Problem 10**

$$-3\sqrt{27} - \sqrt{27} =$$

now apply the distributive property

$$(-3 - 1)\sqrt{27} = -4\sqrt{27} =$$

$$-4\sqrt{3} \cdot \sqrt{9} =$$

$$-4\sqrt{3} \cdot 3 =$$

$$-\mathbf{12}\sqrt{\mathbf{3}}$$

Problems

1	$2\sqrt{2} - \sqrt{18}$	**11**	$-\sqrt{2} - 3\sqrt{5} - 2\sqrt{5}$
2	$-2\sqrt{3} - 2\sqrt{12}$	**12**	$-2\sqrt{5} + 3\sqrt{5} - \sqrt{6}$
3	$2\sqrt{18} + 3\sqrt{18}$	**13**	$-\sqrt{3} + 2\sqrt{5} - 2\sqrt{5}$
4	$3\sqrt{2} - \sqrt{2}$	**14**	$3\sqrt{5} + 2\sqrt{5} - \sqrt{6}$
5	$3\sqrt{8} - 2\sqrt{18}$	**15**	$3\sqrt{5} + 2\sqrt{2} - \sqrt{2}$
6	$-\sqrt{20} - 3\sqrt{45}$	**16**	$2\sqrt{3} + 2\sqrt{5} - \sqrt{3}$
7	$3\sqrt{54} - \sqrt{54}$	**17**	$-2\sqrt{3} + 3\sqrt{3} + 3\sqrt{3}$
8	$-\sqrt{54} - \sqrt{6}$	**18**	$2\sqrt{3} - 2\sqrt{5} + 3\sqrt{3}$
9	$-2\sqrt{45} - 2\sqrt{5}$	**19**	$-\sqrt{6} + 3\sqrt{6} + 3\sqrt{6}$
10	$-3\sqrt{27} - \sqrt{27}$	**20**	$3\sqrt{3} - 2\sqrt{6} + 2\sqrt{3}$

Dividing

Multiplication property of square roots:

$$\sqrt{ab} = \sqrt{a} \cdot \sqrt{b} \quad if \ a \ and \ b \ are \ \geq 0$$

Division property

$$\sqrt{\frac{a}{b}} = \frac{\sqrt{a}}{\sqrt{b}} \quad if \ a \geq o \ and \ b > 0$$

How to Solve

Example 1:

$$\frac{4\sqrt{14}}{8\sqrt{2}} =$$

$$\frac{4}{8} \cdot \frac{\sqrt{14}}{\sqrt{2}} = \frac{1}{2} \cdot \sqrt{\frac{14}{2}} = \frac{1}{2} \cdot \sqrt{7} = \frac{\sqrt{7}}{2}$$

Example 2:

$$\frac{1}{2 \cdot \sqrt{3x}}$$

Multiply by

$$\frac{\sqrt{3x}}{\sqrt{3x}}$$

so

$$\frac{1}{2 \cdot \sqrt{3x}} \cdot \frac{\sqrt{3x}}{\sqrt{3x}} =$$

$$\frac{\sqrt{3x}}{2\sqrt{3^2 \cdot x^2}}$$

$$\frac{\sqrt{3x}}{2 \cdot 3x} = \frac{\sqrt{3x}}{6x}$$

Example 3:

$$\frac{1}{\sqrt{3} - \sqrt{2}} =$$

multiply two conjugates binomials

$$\frac{1}{(\sqrt{3} - \sqrt{2})} \cdot \frac{(\sqrt{3} + \sqrt{2})}{(\sqrt{3} + \sqrt{2})}$$

$$\frac{\sqrt{3} + \sqrt{2}}{\sqrt{9} + \sqrt{6} - \sqrt{6} - \sqrt{4}} =$$

$$\frac{\sqrt{3} + \sqrt{2}}{3 - 2} = \frac{\sqrt{3} + \sqrt{2}}{1} = \sqrt{3} + \sqrt{2}$$

Problems Solved

1) **Problem 18**

$$\frac{5}{5 - 5\sqrt{5}} =$$

$$\frac{5}{5 - 5\sqrt{5}} \cdot \frac{\left(5 + 5\sqrt{5}\right)}{\left(5 + 5\sqrt{5}\right)} = \frac{25 + 25\sqrt{5}}{5^2 - \left(5\sqrt{5}\right)^2} = \frac{25 + 25\sqrt{5}}{25 - 25.5} =$$

$$\frac{25(1 + \sqrt{5})}{25 - 125} = \frac{25(1 + \sqrt{5})}{-100} = \frac{1 + \sqrt{5}}{-4} = \frac{-1 - \sqrt{5}}{4}$$

2) **Problem 10**

$$\frac{2}{3 - \sqrt{5}} =$$

$$\frac{2}{3 - \sqrt{5}} \cdot \frac{\left(3 + \sqrt{5}\right)}{\left(3 + \sqrt{5}\right)} = \frac{6 + 2\sqrt{5}}{3^2 - \left(\sqrt{5}\right)^2} = \frac{2\left(3 + \sqrt{5}\right)}{9 - 5} =$$

$$\frac{2(3 + \sqrt{5})}{4} = \frac{3 + \sqrt{5}}{2}$$

Problems

1	$-\dfrac{5\sqrt{3}}{4-\sqrt{3}}$	**11**	$-\dfrac{1}{3-\sqrt{2}}$
2	$-\dfrac{4}{2-2\sqrt{5}}$	**12**	$\dfrac{4}{4-\sqrt{3}}$
3	$\dfrac{3}{2-4\sqrt{3}}$	**13**	$\dfrac{\sqrt{2}}{-3-2\sqrt{2}}$
4	$-\dfrac{1}{5\sqrt{5}+2\sqrt{3}}$	**14**	$\dfrac{2\sqrt{2}}{2-\sqrt{3}}$
5	$\dfrac{2}{5+\sqrt{3}}$	**15**	$-\dfrac{1}{-4+\sqrt{2}}$
6	$\dfrac{5\sqrt{2}}{-3+\sqrt{5}}$	**16**	$\dfrac{2}{\sqrt{5}+2\sqrt{3}}$
7	$\dfrac{5}{2-3\sqrt{5}}$	**17**	$\dfrac{4}{2+3\sqrt{2}}$
8	$\dfrac{\sqrt{3}}{5+4\sqrt{5}}$	**18**	$\dfrac{5}{5-5\sqrt{5}}$
9	$\dfrac{\sqrt{5}}{3+2\sqrt{3}}$	**19**	$\dfrac{2}{-1+2\sqrt{3}}$
10	$\dfrac{2}{3-\sqrt{5}}$	**20**	$\dfrac{5}{4+\sqrt{3}}$

Multiplying

Multiplication property of square roots:

$$\sqrt{ab} = \sqrt{a} \cdot \sqrt{b} \quad \textit{if a and b are} \geq 0$$

The operator $\geq$ means greater or equal than

How to Solve

Example 1:

$$\sqrt{2} \cdot \sqrt{6}$$

$$\sqrt{2 \cdot 6} =$$

$$\sqrt{12} =$$

$$\sqrt{4 \cdot 3} =$$

$$\sqrt{2^2 \cdot 3} =$$

$$2\sqrt{3}$$

Example 2:

$$\sqrt{2}\left(\sqrt{8} + \sqrt{20}\right) =$$

$$\sqrt{2}\left(\sqrt{2 \cdot 4} + \sqrt{20}\right) =$$

$$\sqrt{2}(2\sqrt{2} + \sqrt{4} \cdot \sqrt{5}) =$$

$$\sqrt{2}\left(2\sqrt{2} + 2\sqrt{5}\right) =$$

$$4 + 2\sqrt{5} \cdot \sqrt{2} =$$

$$4 + 2\sqrt{10}$$

Problems Solved

1) **Problem 17**

$$\sqrt{3}(3 + \sqrt{3})$$
$$\sqrt{3}(3 + \sqrt{3}) = 3\sqrt{3} + \sqrt{9} = \mathbf{3\sqrt{3} + 3}$$

2) **Problem 16**

$$\sqrt{5}(\sqrt{5} + 5)$$
$$\sqrt{5}(\sqrt{5} + 5) = \sqrt{25} + 5\sqrt{5} = \mathbf{5 + 5\sqrt{5}}$$

3) **Problem 1**

$$\sqrt{4} \cdot \sqrt{2}$$
$$\sqrt{4} \cdot \sqrt{2} = \mathbf{2\sqrt{2}}$$

4) **Problem 6**

$$3\sqrt{15} \cdot 3\sqrt{10}$$
$$3\sqrt{15} \cdot 3\sqrt{10} = 9\sqrt{150} =$$

$$factors\ of\ 150:\ 5,5,3,2,1$$

$$9\sqrt{5 \cdot 5 \cdot 3 \cdot 2 \cdot 1} = 9\sqrt{25 \cdot 6} = 9 \cdot 5\sqrt{6} = \mathbf{45\sqrt{6}}$$

5) **Problem 19**

$$\sqrt{6} \cdot (2\sqrt{2} + \sqrt{5})$$
$$\sqrt{6} \cdot (2\sqrt{2} + \sqrt{5}) = 2\sqrt{12} + \sqrt{30} =$$
$$2\sqrt{3 \cdot 4} + \sqrt{30} = \mathbf{4\sqrt{3} + \sqrt{30}}$$

Problems

1	$\sqrt{4} \cdot \sqrt{2}$	11	$\sqrt{5}(\sqrt{5} + 3)$
2	$\sqrt{25} \cdot \sqrt{15}$	12	$-\sqrt{5}(\sqrt{10} + 5)$
3	$\sqrt{15} \cdot \sqrt{20}$	13	$\sqrt{3}(\sqrt{3} + 3)$
4	$\sqrt{5} \cdot \sqrt{3}$	14	$-4\sqrt{10}(\sqrt{2} + 5)$
5	$\sqrt{6} \cdot \sqrt{3}$	15	$5\sqrt{5}(3 - \sqrt{5})$
6	$3\sqrt{15} \cdot 3\sqrt{10}$	16	$\sqrt{5}(\sqrt{5} + 5)$
7	$-4\sqrt{3} \cdot 5\sqrt{5}$	17	$\sqrt{3}(3 + \sqrt{3})$
8	$4\sqrt{5} \cdot 5\sqrt{5}$	18	$-5\sqrt{5}(3 - 4\sqrt{6})$
9	$5\sqrt{20} \cdot -5\sqrt{20}$	19	$\sqrt{6}(2\sqrt{2} + \sqrt{5})$
10	$-5\sqrt{8} \cdot -2\sqrt{6}$	20	$-\sqrt{3}(5 + \sqrt{6})$

Simplify Radical Expressions

How to Solve

Example 1:

$$\sqrt{27}$$

factorize the radicand $27: 3 \cdot 3 \cdot 3$

$$\sqrt{3 \cdot 3 \cdot 3} = \sqrt{9 \cdot 3} =$$

$$\sqrt{9} = 3, so$$

$$\mathbf{3\sqrt{3}}$$

Example 2:

$$\sqrt{75}$$

$$\sqrt{75} = 5\sqrt{5 \cdot 5 \cdot 3} = 5\sqrt{25 \cdot 3} =$$

$$\sqrt{25} = 5, so$$

$$5 \cdot 5\sqrt{3} = \mathbf{25\sqrt{3}}$$

Example 3:

$$\sqrt{108}$$

$$\sqrt{108} = \sqrt{3 \cdot 3 \cdot 3 \cdot 2 \cdot 2} = \sqrt{3 \cdot 9 \cdot 4} =$$

$$\sqrt{9} = 3 \quad and \quad \sqrt{4} = 2, so$$

$$3 \cdot 2\sqrt{3} = \mathbf{6\sqrt{3}}$$

Example 4:

$$\sqrt{48}$$
$$\sqrt{48} = \sqrt{2 \cdot 2 \cdot 2 \cdot 2 \cdot 3} =$$
$$\sqrt{4 \cdot 4 \cdot 3} = 2 \cdot 2\sqrt{3} = \mathbf{4\sqrt{3}}$$

Example 5:

$$3\sqrt{20}$$
$$3\sqrt{20} = 3\sqrt{4 \cdot 5} = 3 \cdot 2\sqrt{5} = \mathbf{6\sqrt{5}}$$

Example 6:

$$12\sqrt{18}$$

$$12\sqrt{18} = 12\sqrt{9 \cdot 2} = 12 \cdot 3\sqrt{2} = \mathbf{36\sqrt{2}}$$

Problems

1	$-6\sqrt{144x}$	**11**	$\sqrt{20x}$
2	$-6\sqrt{448x}$	**12**	$\sqrt{20b^2}$
3	$-\sqrt{128n^2}$	**13**	$\sqrt{32n^2}$
4	$-5\sqrt{75k^3}$	**14**	$\sqrt{30n}$
5	$5\sqrt{80n^3}$	**15**	$\sqrt{16m^2}$
6	$10\sqrt{30}$	**16**	$-10\sqrt{200p^4}$
7	$-6\sqrt{210n}$	**17**	$6\sqrt{42}$
8	$9\sqrt{125x^2}$	**18**	$-10\sqrt{105}$
9	$10\sqrt{448x}$	**19**	$-2\sqrt{18x^2}$
10	$-3\sqrt{500b^4}$	**20**	$4\sqrt{210a}$

RATIONAL EXPRESSIONS

An expression that is rational is simply an expression which has a polynomial. These expressions can be simplified much like fractions can. In order to simplify a rational expression, it is first necessary to determine common factors between the expression's numerators and its denominators. After this, we must then remove these by rewriting the expression, this time setting it equal to 1.

Solving Rational Expressions

A rational equation is defined as an equation which contains at least one rational expression.

Adding and Subtracting

- If the two or more rational expressions that we are trying to add or subtract together have the same denominators, then all that we have to do is add or subtract their numerators.

$$\frac{a}{c} + \frac{b}{c} = \frac{a+b}{c}$$

$$\frac{a}{c} - \frac{b}{c} = \frac{a-b}{c}$$

$$\frac{x}{x-1} + \frac{2-x}{x-1} = \frac{x+2-x}{x-1} = \frac{2}{x-1}$$

- When the denominators of the expressions that you were trying to add or subtract are not the same, then it becomes necessary to find a common denominator among them. The simplest and most common way of doing this is by simply multiplying the denominators together, though this method usually produces complex computations and requires a lot of simplifying afterward. It is nevertheless the best method to use when you are not sure about the computation of the denominators of an expression. The best possible denominator that we can find in these situations, as far as simplifying is concerned, is the LCM. The least common denominator or least common multiple (LCM), is the smallest common multiple between two or more denominators.

Example:

$$\frac{5}{6} - \frac{3}{4}$$

The LCM is 12, so

$$\frac{5}{6} - \frac{3}{4} = \frac{5(2)}{6(2)} - \frac{3(3)}{4(3)} = \frac{10}{12} - \frac{9}{12} = \frac{10-9}{12} = \frac{1}{12}$$

Remember to find LCM:

1) Factors all the denominators
2) Write each factor that appears at least once in any denominators.
3) For each factor in the previous step, write the largest power that occurs in all the denominators containing the factor.
4) The LCM is the product of all the factors in the previous step

How to Solve

Example 1:

$$\frac{4}{6x^2} - \frac{1}{3x^5} + \frac{5}{2x^3}$$

$$Factors\ of\ 6x^2: \quad 6 \cdot x \cdot x$$

$$Factors\ of\ 3x^2: \quad 3 \cdot x \cdot x$$

$$Factors\ of\ 2x^3: \quad 2 \cdot x \cdot x \cdot x$$

$$LCM = 6x^5$$

LCM is $6x^5$, because 6 is the term containing 6, 3 and 2 x^5 is the largest power that occurs on all the x^n denominator

$$\frac{4}{6x^2} - \frac{1}{3x^5} + \frac{5}{2x^3} =$$

$$\frac{4(x^3)}{6x^2(x^3)} - \frac{1(2)}{3x^5(2)} + \frac{5(3x)^2}{2x^3(3x)^2} =$$

$$\frac{4x^3}{6x^5} - \frac{2}{6x^5} + \frac{15x^2}{6x^5} =$$

$$\frac{4x^3 - 2 + 15^2}{6^5}$$

Example 2:

$$\frac{2}{z+1} - \frac{z-1}{z+2}$$

$$LCM = (z+1)(z+2)$$

$$\frac{2(z+2)}{(z+1)(z+2)} - \frac{(z-1)(z+1)}{(z+1)(z+2)} =$$

$$\frac{2z+4-z^2+1}{(z+1)(z+2)} =$$

$$\frac{-z^2+2z+5}{(z+1)(z+2)}$$

Problems Solved

1) **Problem 1**

$$\frac{3n}{6} + \frac{n+1}{n-3}$$

$$LCM = 6(n-3)$$

now for each fraction you multiply the numerator by the division between LCM and the denominator of the fraction, so for the first fraction you divide the LCM by the denominator 6, so:

$$6(n-3) \div 6 = (n-3)$$

now you multiply the numerator $3n$ by $(n-3)$.

Repeat these procedure for each fraction of the expression

$$\frac{3n}{6} + \frac{n+1}{n-3} = \frac{3n(n-3) + 6(n+1)}{6(n-3)} =$$

$$\frac{3n^2 - 9n + 6n + 6}{6(n-3)} = \frac{3n^2 - 3n + 6}{6(n-3)} = \frac{3(n^2 - n + 2)}{6(n-3)} =$$

now simplify the fraction between 3 and 6

by dividing by 2, so:

$$= \frac{n^2 - n + 2}{2(n-3)}$$

2) Problem 8

$$\frac{6}{r-1} - \frac{3r}{3r-2} =$$

$$LCM = (r-1)(3r-2)$$

$$\frac{6(3r-2) - 3r(r-1)}{(r-1)(3r-2)} = \frac{18r - 12 - 3r^2 + 3r}{(r-1)(3r-2)} =$$

$$\frac{-3r^2 + 21r - 12}{(r-1)(3r-2)}$$

3) Problem 5

$$\frac{3x-2}{10x+8} + \frac{4x}{2x}$$

$$LCM = (10x+8)(2x)$$

$$\frac{\cancel{2x}(3x-2)}{(10x+8)(\cancel{2x})} + \frac{4x(\cancel{10x+8})}{(\cancel{10x+8})(2x)}$$

$$\frac{3x-2}{(10x+8)} + 2$$

$$\frac{(3x-2) + 2(10x+8)}{10x+8}$$

$$\frac{3x - 2 + 20x + 16}{2(5x+4)} = \frac{23x + 14}{2(5x+4)}$$

Problems

1	$\dfrac{3n}{6} + \dfrac{n+1}{n-3}$	**11**	$\dfrac{3x+2}{2x^2+10x} + \dfrac{6}{2x}$
2	$\dfrac{2}{3} + \dfrac{4x}{3x^2+18x}$	**12**	$\dfrac{k+1}{3k+9} - \dfrac{6}{3k}$
3	$\dfrac{5}{3r-1} - \dfrac{6}{3}$	**13**	$\dfrac{a+6}{a^2-4a+3} - 4a$
4	$\dfrac{6}{6m} + \dfrac{m+5}{m-3}$	**14**	$\dfrac{6x}{x-4} + \dfrac{x+4}{6}$
5	$\dfrac{3x-2}{10x+8} + \dfrac{4x}{2x}$	**15**	$\dfrac{x+2}{6x^2+10x} + \dfrac{x+6}{3}$
6	$\dfrac{4}{5x^2} + \dfrac{2x}{3x+2}$	**16**	$\dfrac{4}{3} - \dfrac{2b}{b+5}$
7	$\dfrac{5n}{3} - \dfrac{6}{3n-15}$		
8	$\dfrac{6}{r-1} - \dfrac{3r}{3r-2}$		
9	$\dfrac{5x}{x+2} + \dfrac{5x}{x-1}$		
10	$\dfrac{3r}{r-5} - \dfrac{3r}{r-2}$		

Dividing

Please see previous chapter "The Remainder Theorem Method in the Polynomials chapter.

How to Solve

Given a polynomial P(x) with degree at least 1 and any number r there is a polynomial **Q(x), called the quotient**, with degree one less than the degree of P(x) and a number **R, called the remainder**
so that

$$P(x) = (x - r)\, Q(x) + R$$

Problems

Please see previous chapter "The Remainder Theorem Method in the Polynomials chapter.

Multiplying

How to Solve

1) **Problem 1**

$$\frac{v+5}{9} \cdot \frac{v^2 - 4v - 12}{v+5} =$$

the factors of $v^2 - 4v - 12 = (v-6)(v+2)$

$$\frac{(v+5)}{9} \cdot \frac{(v-6)(v+2)}{(v+5)} =$$

$$\frac{(v-6)(v+2)}{9}$$

2) **Problem 2**

$$\frac{1}{a+1} \cdot \frac{2a-2}{a^2 - 10a + 9} =$$

the factor of $a^2 - 10a + 9$ is $(a-9)(a-1)$

$$\frac{2a-2}{(a+1)(a-9)(a-1)} =$$

$$\frac{2(a-1)}{(a+1)(a-9)(a-1)} =$$

$$\frac{2(a-1)}{(a+1)(a-9)(a-1)} =$$

$$\frac{2}{(a+1)(a-9)}$$

3) **Problem 9**

$$\frac{a^2 - 36}{a + 6} \cdot \frac{1}{a - 9}$$

$a^2 - 36 \ can \ be \ factorized \ in \ (a - 6)(a + 6)$

$$\frac{(a - 6)\cancel{(a + 6)}}{\cancel{(a + 6)}(a - 9)} =$$

$$\frac{a - 6}{a - 9}$$

4) **Problem 6**

$$\frac{1}{k - 5} \cdot \frac{10k - 50}{k^2 - 5k - 24}$$

$10k - 50 \ can \ be \ factorized \ in \ 10(k - 5)$

$k^2 - 5k - 24 \ can \ be \ factorized \ in \ (x + 3)(x - 8)$

$$\frac{10\cancel{(k - 5)}}{\cancel{(k - 5)}(k + 3)(k - 8)} =$$

$$\frac{10}{(k + 3)(k - 8)}$$

5) **Problem 3**

$$\frac{3}{2b - 18} \cdot \frac{b^2 - 5b - 36}{b + 4} =$$

$check \ which \ factorizations \ we \ can \ do$

$$b^2 - 5b - 36 = (b - 9)(b + 4)$$

$$2b - 18 = 2(b - 9)$$

$$\frac{3}{2\cancel{(b - 9)}} \cdot \frac{\cancel{(b - 9)}\cancel{(b + 4)}}{\cancel{(b + 4)}} = \frac{3}{2}$$

Problems

1	$\dfrac{v+5}{9} \cdot \dfrac{v^2-4v-12}{v+5}$
2	$\dfrac{1}{a+1} \cdot \dfrac{2a-2}{a^2-10a+9}$
3	$\dfrac{3}{2b-18} \cdot \dfrac{b^2-5b-36}{b+4}$
4	$\dfrac{x^2-4x-32}{x+9} \cdot \dfrac{1}{x+4}$
5	$\dfrac{1}{x+4} \cdot \dfrac{x^2+12x+32}{9x}$
6	$\dfrac{1}{k-5} \cdot \dfrac{10k-50}{k^2-5k-24}$
7	$\dfrac{10p-8}{2} \cdot \dfrac{p+4}{40p-32}$

8	$$\frac{x^2 - 10x + 25}{x - 5} \cdot \frac{1}{x + 5}$$
9	$$\frac{a^2 - 36}{a + 6} \cdot \frac{1}{a - 9}$$
10	$$\frac{x - 3}{6} \cdot \frac{6x + 30}{x + 5}$$
11	$$\frac{2n^2 - 18n}{2n} \cdot \frac{1}{8n^2}$$
12	$$\frac{x + 6}{x + 3} \cdot \frac{x^2 + 4x + 3}{2x^3 + 2x^2}$$
13	$$\frac{9r - 9}{1 - r} \cdot \frac{r - 3}{9}$$
14	$$\frac{p^2 - 5p - 50}{p - 7} \cdot \frac{7 - p}{p^2 - 7p - 30}$$
15	$$\frac{9}{x^2 + 6x - 27} \cdot \frac{x^2 - 11x + 24}{9}$$

16	$\dfrac{n^2 - n - 90}{n + 9} \cdot \dfrac{1}{n + 4}$
17	$\dfrac{1}{x - 8} \cdot \dfrac{x^2 - 17x + 72}{9x^2 - 27x}$
18	$\dfrac{p^2 - 3p - 18}{2} \cdot \dfrac{p - 1}{p^2 + 2p - 3}$
19	$\dfrac{1}{x - 7} \cdot \dfrac{x^2 - 6x - 7}{7}$
20	$\dfrac{x - 3}{3 - x} \cdot \dfrac{x^2 + 6x - 27}{x + 9}$

Simplifying

A rational expression is only considered simplified when its numerators and its denominators have no factors in common.

Simplifying rational expressions can be carried out in the same manner that simplifying numerical fractions is always carried out.

For example, one simplified version of 6/8 is ¾. To arrive at this simplified version of the fraction, we had to cancel out the common factor of 2 from both numerators and denominators of both fractions:

$$\frac{8}{6} = \frac{2 \cdot 4}{2 \cdot 3} = \frac{4}{3}$$

How to Solve

factorize numerator and denominator then simplify

$$\frac{x^2 - 2x - 8}{x^2 - 9x + 20} = \frac{(x-4)(x+2)}{(x-5)(x-4)} = \frac{(x+2)}{(x-5)}$$

Problems Solved

1) **Problem 1**

$$\frac{n+4}{n^2 + 9n + 20}$$

$$\frac{n+4}{n^2 + 9n + 20} = \frac{(n+4)}{(n+5)(n+4)} = \frac{1}{n+5}$$

2) **Problem 6**

$$\frac{15m}{25m - 25}$$

$$\frac{15m}{25m - 25} = \frac{15m}{25(m-1)} = \frac{3 \cdot 5m}{5 \cdot 5(m-1)} = \frac{3m}{5(m-1)}$$

3) **Problem 13**

$$\frac{12x^2 + 8x}{16x}$$

$$\frac{12x^2 + 8x}{16x} = \frac{4x(3x+2)}{4 \cdot 4x} = \frac{3x+2}{4}$$

Problems

1	$\dfrac{n+4}{n^2+9n+20}$	**11**	$\dfrac{r^2+5r+4}{r+4}$
2	$\dfrac{n+1}{n^2+3n+2}$	**12**	$\dfrac{5n^2-10n}{n-2}$
3	$\dfrac{r^2+r-12}{r-3}$	**13**	$\dfrac{12x^2+8x}{16x}$
4	$\dfrac{r^2+8r+15}{r+5}$	**14**	$\dfrac{v-5}{3v^2-15v}$
5	$\dfrac{6x-4}{8}$	**15**	$\dfrac{a+3}{a^2+4a+3}$
6	$\dfrac{15m}{25m-25}$	**16**	$\dfrac{16v}{20v+4}$
7	$\dfrac{a^2+a-2}{a-1}$	**17**	$\dfrac{a-4}{3a-12}$
8	$\dfrac{12}{9x+6}$	**18**	$\dfrac{8x+8}{16x}$
9	$\dfrac{p-2}{p^2-4p+4}$	**19**	$\dfrac{x-3}{5x-15}$
10	$\dfrac{n^2-4}{n+2}$	**20**	$\dfrac{9n}{6n^2+6n}$

ONE STEP EQUATION

One step equation is, as their title suggests, the simplest and most straightforward types of equations that algebra includes. The main skill to develop in solving these is to know how to perform the inverse of the operation within the equation. If the equation features addition, then subtraction needs to be used to isolate the variable, and vice versa. If the equation features multiplication, then division needs to be used to isolate the variable, and again, vice versa. We will now solve for examples of equations which use these various functions below:

How to Solve

1) To solve for $5 + x = 10$ we would need to subtract both sides by 5 in order to isolate the variable:

$$5 + x - 5 = 10 - 5$$
$$x = 5$$

2) To solve for $x - 3 = 10$, we would need to add both sides by 3 in order to isolate the variable:

$$x - 3 + 3 = 10 + 3$$
$$x = 13$$

3) To solve for $3x = 18$, we would need to divide both sides by 3 so we can isolate the variable:

$$\frac{3x}{3} = \frac{18}{3}$$

$$x = 6$$

4) To solve for $\frac{x}{2} = 10$, we would need to multiply 2 by both sides in order to isolate the variable:

$$\left(\frac{x}{2}\right)2 = (10)2$$

$$x = 20$$

Quick Method:

Below some problem solved by means oa a quick method, that I propose:

1)   Example 1:

$$x + 10 = 12$$

You must isolate the x at left side of equal, moving the number on the right side of equal. To do this you have to change the sign

$$x = -10 + 12 = 2$$

2)   Example 2:

You must isolate the r at left side of equal, moving the denominator on the right side of equal. To do this you have to multiply the denominator

$$\frac{r}{4} = 24$$
$$r = 24 \cdot 4 = 96$$

Problems Solved

1)   Problem 9

You must isolate the p at left side of equal, moving the coefficient on the right side of equal. To do this you have to divide the number on the right by the coefficient on the left side.

$$\frac{5}{13}p = \frac{15}{26}$$

now simplify with criss − cross

$$p = \frac{15}{26} \div \frac{5}{13} = \frac{\cancel{15}}{\cancel{26}} \cdot \frac{\cancel{13}}{\cancel{5}} =$$

$$\frac{15 \div 5}{26 \div 13} \cdot \frac{13 \div 13}{5 \div 5} = \frac{3}{2}$$

$$resut = \frac{3}{2}$$

2) **Problem 15**

$$p - -1 = -13$$
$$p + 1 = -13$$
$$p = -1 - 13$$
$$\boldsymbol{p = -14}$$

3) **Problem 8:**

$$n - 1 = \frac{7}{6}$$
$$n = 1 + \frac{7}{6}$$
$$\boldsymbol{n = \frac{13}{6}}$$

4) **Problem 10:**

$$x - \frac{10}{3} = \frac{71}{30}$$
$$x = \frac{71}{30} + \frac{10}{3} =$$
$$x = \frac{71 + 100}{30} = \frac{171}{30} = \boldsymbol{\frac{57}{10}}$$

Because the fraction can be simplified by dividing both by 3

5) **Problem 8:**

$$n - 1 = \frac{7}{6}$$
$$n = \frac{7}{6} + 1$$
$$n = \frac{7 + 6}{6}$$
$$\boldsymbol{n = \frac{13}{6}}$$

6) Problem 18:

$$\frac{161}{75} = \frac{42}{25} + k$$

$$\frac{161}{75} - \frac{42}{25} = k$$

$$\frac{161 - 126}{75} = k$$

$$\frac{35}{75} = k$$

we can simplify by dividing both numbers by 5,

$$k = \frac{7}{15}$$

7) Problem 16:

$$\frac{26}{9} a = -\frac{52}{135}$$

$$a = -\frac{52}{135} \cdot \frac{9}{26} =$$

now we can do the simplifications, so

$$-\frac{52 \div 26}{135 \div 3} \cdot \frac{9 \div 3}{26 \div 26} = -\frac{2}{45} \cdot 3 = -\frac{6}{45} =$$

$$-\frac{6 \div 3}{45 \div 3} = -\frac{2}{15}$$

$$a = -\frac{2}{15}$$

Problems

1	$x + 10 = 12$	**11**	$45 = 23 - v$
2	$x - 5 = -2$	**12**	$22 = \dfrac{n}{25}$
3	$-5n = -45$	**13**	$7 = 22 - x$
4	$b - 4 = -11$	**14**	$\dfrac{r}{4} = 24$
5	$-8x = -48$	**15**	$p - -1 = -13$
6	$2p = \dfrac{9}{4}$	**16**	$\dfrac{26}{9}a = -\dfrac{52}{135}$
7	$a + \dfrac{2}{7} = -3$	**17**	$\dfrac{1719}{77} = \dfrac{167}{11} + p$
8	$n - 1 = \dfrac{7}{6}$	**18**	$\dfrac{161}{75} = \dfrac{42}{25} + k$
9	$\dfrac{5}{13}p = \dfrac{15}{26}$	**19**	$a + \dfrac{45}{29} = \dfrac{1586}{667}$
10	$x - \dfrac{10}{3} = \dfrac{71}{30}$	**20**	$-\dfrac{17}{18}k = \dfrac{1139}{360}$

MULTI STEP EQUATION

Equations which require more than one step to solve

How to Solve

1) **Problem 1:**

$$-8(2b + 3) = 104$$
$$-16b - 24 = 104$$
$$-16b = 24 + 104$$
$$-16b = 128$$
$$-b = \frac{128}{16}$$
$$b = -\frac{128}{16}$$
$$b = -8$$

2) **Problem 5:**

$$3(5x - 6) = -138$$
$$15x - 18 = -138$$
$$15x = 18 - 138$$
$$15x = -120$$
$$x = \frac{-120}{15} = -8$$

3) Problem 20:

$$\frac{13}{2}\left(\frac{7}{3}x + \frac{5}{3}\right) = -\frac{221}{2}$$

$$\frac{91}{6}x + \frac{65}{6} = -\frac{221}{2}$$

$$\frac{91}{6}x = -\frac{65}{6} - \frac{221}{2}$$

$$\frac{91}{6}x = \frac{-65 - 663}{6}$$

$$\frac{91}{6}x = -\frac{728}{6}$$

Simplify the fraction (divisible by 2)

$$-\frac{728}{6} = -\frac{364}{3}$$

$$\frac{91}{6}x = -\frac{364}{3}$$

$$x = -\frac{364}{3} \cdot \frac{6}{91} = -\frac{728}{91} = -8$$

Solution $x = -\textbf{8}$

4) Problem 10:

$$7(4b - 5) + 8b = 217$$

$$28b - 35 + 8b = 217$$

$$36b = 35 + 217$$

$$36b = 252$$

$$b = \frac{252}{36} = 7$$

solution $\boldsymbol{b} = \textbf{7}$

5)   Problem 11:

$$7\left(\frac{5}{2}x - \frac{1}{3}\right) + \frac{9}{2}x = \frac{521}{3}$$

$$\frac{35}{2}x - \frac{7}{3} + \frac{9}{2}x = \frac{521}{3}$$

$$\left(\frac{35}{2} + \frac{9}{2}\right)x - \frac{7}{3} = \frac{521}{3}$$

$$\frac{44}{2}x = \frac{7}{3} + \frac{521}{3}$$

$$22x = \frac{528}{3}$$

$$x = \frac{528}{3} \cdot \frac{1}{22} = \frac{528}{66} = 8$$

solution $x = 8$

6)   Problem 18:

$$\frac{23}{5}\left(\frac{21}{5}m - 2\right) = \frac{437}{5}$$

$$\frac{483}{25}m - \frac{46}{5} = \frac{437}{5}$$

$$\frac{483}{25}m = \frac{437}{5} + \frac{46}{5}$$

$$\frac{483}{25}m = \frac{483}{5}$$

$$m = \frac{483}{5} \cdot \frac{25}{483} = \frac{483 \div 483}{5 \div 5} \cdot \frac{25 \div 5}{483 \div 483} = 5$$

Solution $m = 5$

7) Problem 3:

$$5(-1 + 4n) = 115$$

$$-5 + 20n = 115$$

$$20n = 115 + 5$$

$$20n = 120$$

$$n = \frac{120}{20} = 6$$

Solution $\boldsymbol{n = 6}$

Problems

1	$-8(2b+3) = 104$	**11**	$7\left(\dfrac{5}{2}x - \dfrac{1}{3}\right) + \dfrac{9}{2}x = \dfrac{521}{3}$
2	$6(3x+5) = 102$	**12**	$-\dfrac{20}{7}\left(\dfrac{33}{4}a + \dfrac{2}{7}\right) + \dfrac{12}{7}a = \dfrac{4091}{49}$
3	$5(-1+4n) = 115$	**13**	$\dfrac{1412}{15} = 8\left(\dfrac{7}{6}a + \dfrac{18}{5}\right)$
4	$4(1-5x) = 84$	**14**	$-\dfrac{38497}{400} = -\dfrac{11}{8} - \dfrac{39}{5}\left(\dfrac{7}{2}x - \dfrac{7}{5}\right)$
5	$3(5x-6) = -138$	**15**	$\dfrac{3}{5} - \dfrac{17}{5}\left(\dfrac{27}{7}n + \dfrac{10}{7}\right) = -\dfrac{2903}{35}$
6	$-5(8k-1) = 285$	**16**	$-6\left(\dfrac{41}{8}x - \dfrac{7}{4}\right) = -133$
7	$3 - 8(7+x) = -109$	**17**	$-8a - \dfrac{3}{2}\left(a + \dfrac{29}{6}\right) = -\dfrac{333}{4}$
8	$-5(7-5n) = -235$	**18**	$\dfrac{23}{5}\left(\dfrac{21}{5}m - 2\right) = \dfrac{437}{5}$
9	$4(6+4r) = 104$	**19**	$\dfrac{10}{3}p + 5\left(\dfrac{7}{2}p + \dfrac{31}{8}\right) = \dfrac{4465}{24}$
10	$7(4b-5) + 8b = 217$	**20**	$\dfrac{13}{2}\left(\dfrac{7}{3}x + \dfrac{5}{3}\right) = -\dfrac{221}{2}$

EQUATIONS WITH ABSOLUTE VALUE

These equations contain absolute values

How to Solve

Example 1:

$$|2x - 5| = 9$$

$$|2x - 5| = 9$$

$$2x - 5 = -9 \quad or \quad 2x - 5 = 9$$

Now we have two linear equations easy to solve

$$2x = -4 \quad or \quad 2x = 14$$

$$x = -2 \quad or \quad x = 7$$

The solutions are:

$$x = -2 \quad and \quad x = 7$$

Example 2:

$$|5y - 8| = 1$$

$$5y - 8 = -1 \quad or \quad 5y - 8 = 1$$

$$5y = 7 \quad or \quad 5y = 9$$

$$y = \frac{7}{5} \quad or \quad y = \frac{9}{5}$$

The solutions are

$$x = \frac{7}{5} \quad and \quad x = \frac{9}{5}$$

Problems Solved

1) **Problem 11**

$$\left| a - \frac{11}{6} \right| = \frac{28}{15}$$

$$a - \frac{11}{6} = \frac{28}{15} \quad or \quad a - \frac{11}{6} = -\frac{28}{15}$$

now we solve the first equation

$$a = \frac{28}{15} + \frac{11}{6} = \frac{56 + 55}{30} = \frac{111}{30} = \frac{37}{10}$$

$$a = \frac{37}{10}$$

now we solve the second equation

$$a - \frac{11}{6} = -\frac{28}{15}$$

$$a = -\frac{28}{15} + \frac{11}{6}$$

$$a = \frac{-56 + 55}{30} = -\frac{1}{30}$$

$$a = -\frac{1}{30}$$

$$solutions \quad \frac{37}{10}, -\frac{1}{30}$$

2) Problem 14

$$\left|x - \frac{3}{2}\right| = \frac{1}{2}$$

$$x - \frac{3}{2} = \frac{1}{2} \quad or \quad x - \frac{3}{2} = -\frac{1}{2}$$

now we solve the first equation

$$x - \frac{3}{2} = \frac{1}{2}$$

$$x = \frac{1}{2} + \frac{3}{2}$$

$$x = \frac{4}{2} = 2$$

now we solve the second equation

$$x - \frac{3}{2} = -\frac{1}{2}$$

$$x = -\frac{1}{2} + \frac{3}{2}$$

$$x = \frac{2}{2} = 1$$

the solutions are:

2,1

Problems

1	$|x - 8| = 6$	**11**	$\left|a - \dfrac{11}{6}\right| = \dfrac{28}{15}$
2	$|9 + k| = 5$	**12**	$\left|\dfrac{9}{47}n\right| = \dfrac{81}{235}$
3	$\left|\dfrac{n}{7}\right| = 5$	**13**	$\left|-\dfrac{37}{10}x\right| = \dfrac{333}{80}$
4	$\left|\dfrac{b}{2}\right| = 3$	**14**	$\left|x - \dfrac{3}{2}\right| = \dfrac{1}{2}$
5	$\left|\dfrac{n}{10}\right| = 1$	**15**	$\left|-\dfrac{8}{5}b\right| = \dfrac{144}{25}$
6	$\left|\dfrac{n}{6}\right| = 2$	**16**	$\left|x - \dfrac{16}{9}\right| = \dfrac{79}{72}$
7	$|-7 + x| = 6$	**17**	$\left|x - \dfrac{3}{2}\right| = \dfrac{6}{5}$
8	$|4 + a| = 4$	**18**	$\left|\dfrac{5}{4}m\right| = \dfrac{17}{8}$
9	$|r + 5| = 6$	**19**	$\left|\dfrac{2}{5}b\right| = \dfrac{7}{15}$
10	$|n - 1| = 3$	**20**	$\left|p + \dfrac{7}{6}\right| = \dfrac{7}{6}$

QUADRATIC EQUATIONS

Solve by Factoring

How to Solve

Example 1:

Solve by factoring the following Quadratic Equation

$$x^2 + 7x - 30 = 0$$

Remember: to find factory quickly you take the number without variable, in this case -30, then apply the divisibility rule, 30 is divisible by 10, and then by 3, so

$$
\begin{array}{c|c}
30 & 10 \\
3 & 3 \\
\hline
1 &
\end{array}
$$

The factoring in the form $(x \pm z1)(x \pm z2)$
In this case

$$z1 \cdot z2 = -30 \quad so\ z1 = -3\ and\ z2 = 10$$

$$so$$

$$(x - 3)(x + 10) = 0$$

After factoring the quadratic on the left hand side we can rewrite:

$$(x - 3)(x + 10) = 0$$

The only way any equation of the form $A \cdot B = 0$ can be satisfied if $A = 0$ or $B = 0$, so the equation

$$(x - 3)(x + 10) = 0$$

$Can\ be\ satisfied\ if\ (x - 3) = 0\ or\ (x + 10) = 0$

Solving these two simple linear equations the solution is

$$x = 3, \quad and \ -10$$

Example 2:

$$8v^2 - 64v = 0$$

$$8v(v - 8) = 0$$

$Can\ be\ satisfied\ if \quad 8v = 0\ or\ \ v - 8 = 0$

$Solutions:\ v = 0\ \ v = 8$

Example 3:

$$m^2 + 5m - 14 = 0$$

$$(m - 2)(m + 7) = 0$$

$Can\ be\ satisfied\ if$

$(m - 2) = 0 \quad or\ \ (m + 7) = 0$

$Solutions:\ m = 2 \quad m = -7$

Problems

1	$5n^2 + 40n = 0$	**11**	$7k^2 - 56k + 49 = 0$
2	$x^2 + 3x = 0$	**12**	$3p^2 + 15p + 12 = 0$
3	$8v^2 - 64v = 0$	**13**	$n^2 + 8n + 16 = 0$
4	$n^2 - 8n + 15 = 0$	**14**	$3x^2 - 75 = 0$
5	$4v^2 + 12v - 40 = 0$	**15**	$4a^2 + 36a + 72 = 0$
6	$x^2 - 7x - 8 = 0$	**16**	$m^2 + 5m - 14 = 0$
7	$6n^2 - 6n - 180 = 0$	**17**	$x^2 - 10x + 21 = 0$
8	$5n^2 + 55n + 120 = 0$	**18**	$n^2 + 4n = 0$
9	$2x^2 - 16x + 24 = 0$	**19**	$r^2 - 10r + 16 = 0$
10	$8v^2 + 128v + 512 = 0$	**20**	$x^2 + 2x = 0$

Solving by Quadratic Formula

The simplest and most common way of solving the equation

$$ax^2 + bx + c = 0$$

regarding the quantity of x is by factoring our quadratic. This procedure involves setting every factor equal to zero and then solving for each factor from there. There are times, however, at which quadratics will be too messy to factor or they won't factor at all. While factoring may not always be successful or may not always be the best possible course of action for us, the quadratic formula can always find us the solutions that we are looking for. The quadratic formula uses the c, b, and a from the equation $ax^2 + bx + c$ where all these letters are only taken for numbers; these numbers representing the numerical coefficients of the quadratic equation that they are supposed to solve.

How to solve

To solve the quadratic equations use the quadratic formula

$$\frac{-b \pm \sqrt{b^2 - 4ac}}{2a}$$

1) if $b^2 - 4ac > 0$ *then we have two real solutions*

2) if $b^2 - 4ac = 0$ *then we have a double root*

3) if $b^2 - 4ac < 0$ *then we have two complex solutions*

Example 1

$$x^2 + 3x - 10 = 0$$

so

$$a = 1, b = 3, c = -10$$

$$\frac{-b \pm \sqrt{b^2 - 4ac}}{2a} = \frac{-3 \pm \sqrt{3^2 - 4(1)(-10)}}{2(1)} =$$

$$\frac{-3 \pm \sqrt{9 + 40}}{2} = \frac{-3 \pm \sqrt{49}}{2} = \frac{-3 \pm 7}{2} =$$

$$x = \frac{-10}{2} = -5$$

$$x = \frac{4}{2} = 2$$

The solution set is $\{-5, 2\}$

Example 2:

$$x^2 + 4x - 3 = 0$$

so

$$a = 1, \quad b = 4, \quad c = -3$$

$$x = \frac{-b \pm \sqrt{b^2 - 4ac}}{2a}$$

$$\frac{-4 \pm \sqrt{(4)^2 - 4(1)(-3)}}{2(1)} =$$

$$\frac{-4 \pm \sqrt{16 + 12}}{2} =$$

$$\frac{-4 \pm \sqrt{28}}{2} =$$

$$\sqrt{28} = \sqrt{4 \cdot 7} = 2 \cdot \sqrt{7}$$

$$\frac{-4 \pm 2\sqrt{7}}{2} =$$

$$\frac{\cancel{2}(-2 \pm \sqrt{7})}{\cancel{2}} = -2 \pm \sqrt{7}$$

The solution set is $\{-2 + \sqrt{7}, \ | \ -2 - \sqrt{7}\}$

Problems

1	$x^2 - x - 12 = 0$	**11**	$2n^2 - 3n - 3 = 2$
2	$n^2 - 2n + 5 = 0$	**12**	$k^2 - 4k - 8 = 4$
3	$2x^2 + x - 21 = 0$	**13**	$2r^2 + 2r - 21 = 3$
4	$n^2 - 3n + 2 = 0$	**14**	$2v^2 + v - 6 = -3$
5	$2x^2 - 2x - 24 = 0$	**15**	$2x^2 - 3x - 5 = -3$
6	$n^2 + n - 6 = 0$	**16**	$x^2 + 3x - 8 = -4$
7	$2n^2 + 4n - 16 = 0$	**17**	$2p^2 + p - 3 = 3$
8	$v^2 - 3v - 4 = 0$	**18**	$v^2 - 2v - 28 = -4$
9	$2v^2 - 3v + 1 = 0$	**19**	$2x^2 + 5x - 3 = 4$
10	$2x^2 - x - 3 = 0$	**20**	$m^2 - m - 24 = -4$

Solving by taking square roots

Rules

1) *For any number (n) greater than 0, if $x^2 = n$, then*

$$x = \pm\sqrt{n}$$

2) **Square Root Property**

$$a\sqrt{x} \cdot b\sqrt{y} = ab\sqrt{y}$$

3) ***Product Properties***

$$\sqrt{a} \cdot \sqrt{b} = \sqrt{a \cdot b} \quad \textit{where a and b} > 0$$

$$\sqrt{a} \cdot \sqrt{a} = \sqrt{a^2} = a \qquad \textit{where a} > 0$$

$$\sqrt{xy} = \sqrt{x} \cdot \sqrt{y}, \quad \textit{where x, y} \geq 0$$

4) ***Quotient Property***

$$\sqrt{\frac{x}{y}} = \frac{\sqrt{x}}{\sqrt{y}} \quad \textit{where x} \geq 0 \quad \textit{and } y > 0$$

5) **Exponent Property**

$$\left(\sqrt{x}\right)^2 = x$$

6) **Square root of a negative number**

$$\sqrt{-r} = \sqrt{-1 \cdot r} = \sqrt{-1} \cdot \sqrt{r} = i\sqrt{r}$$

i' is the imaginary unit and equals the square root of -1, *so*

$$i = \sqrt{-1}$$

Example:

$$\sqrt{-60} = \sqrt{-1 \cdot 60} = \sqrt{-1 \cdot 2 \cdot 2 \cdot 3 \cdot 5} = \sqrt{-1 \cdot 2^2 \cdot 3 \cdot 5} =$$

$$\sqrt{-1} \cdot \sqrt{2^2} \cdot \sqrt{3 \cdot 5} = i \cdot 2\sqrt{15} = 2i\sqrt{15}$$

7) **Distributive Property**

$$a\sqrt{c} + b\sqrt{c} = (a + b)\sqrt{c}$$

How to solve

Example 1:

$$x^2 = 36$$
$$\sqrt{x^2} = \pm\sqrt{36}$$
$$x = \pm 6$$

Example 2:

$$\sqrt{3} \cdot \sqrt{15} = \sqrt{45}$$

$$simplify\ the\ result\ if\ possible, so$$

$$= \sqrt{9 \cdot 5}$$

Example 3:

$$\sqrt{2} \cdot \sqrt{8} = \sqrt{16} = \sqrt{4^2} = 4$$

Example 4:

$$2x^2 - 16 = 0$$
$$2x^2 = 16$$
$$x^2 = \frac{16}{2} = 8$$
$$x = \pm\sqrt{8}$$
$$x = \pm\sqrt{4} \cdot \sqrt{2}$$
$$x = \pm 2\sqrt{2}$$

Example 5:

$$-5x^2 + 9 = 0$$

$$-5x^2 = -9$$

$$5x^2 = 9$$

$$x^2 = \frac{9}{5}$$

Apply square root

$$x = \pm\sqrt{\frac{9}{5}} =$$

Use the quotient property

$$\pm\frac{\sqrt{9}}{\sqrt{5}} =$$

Simplify the numerator

$$\pm\frac{3}{\sqrt{5}} =$$

Rationalize the denominator

$$\pm\frac{3\sqrt{5}}{5}$$

Example 6:

$$\left(\sqrt{7}\right)^2 = 7$$

Example 7:

$$\frac{\sqrt{24}}{\sqrt{6}} = \sqrt{\frac{24}{6}} = \sqrt{4} = 2$$

Problems Solved

1)   Problem 8

$$b^2 = 68$$

The number 68 is divisible by 2, so you make the prime factorization, do you remember?

$$
\begin{array}{r|l}
68 & 2 \\
34 & 2 \\
17 & 17 \\
1
\end{array}
$$

$$\sqrt{b} = \pm\sqrt{68} = \pm\sqrt{2 \cdot 2 \cdot 17} = \pm\sqrt{4 \cdot 17} =$$

$$\pm\sqrt{4} \cdot \sqrt{17} = \pm 2\sqrt{17}$$

So the solutions are: $2\sqrt{17}, -2\sqrt{17}$

2)   Problem 7

$$r^2 = 50$$

The number 50 is divisible by 5, so you make the prime factorization, do you remember?

$$
\begin{array}{r|l}
50 & 5 \\
10 & 5 \\
2 & 2 \\
1
\end{array}
$$

$$\sqrt{r} = \pm\sqrt{50} = \pm\sqrt{5 \cdot 5 \cdot 2} = \pm\sqrt{25} \cdot \sqrt{2} = \pm 5\sqrt{2}$$

So the solutions are: $5\sqrt{2}, -5\sqrt{2}$

3)   Problem 6

$$r^2 = -21$$

The number 21 is divisible by 3, so you make the prime factorization, do you remember ?

$$21|3$$
$$7|7$$
$$1$$

$$\sqrt{r} = \pm\sqrt{-21} = \pm\sqrt{-3 \cdot 7} = \pm\sqrt{-3} \cdot \sqrt{7} =$$

No solutions

4)   Problem 14

$$x^2 = 90$$

The number 90 is divisible by 10, so you make the prime factorization, do you remember ?

$$90|10$$
$$9|3$$
$$3|3$$
$$1$$

$$\sqrt{x} = \pm\sqrt{90} = \pm\sqrt{10 \cdot 3 \cdot 3} = \pm\sqrt{10} \cdot \sqrt{9} = \pm 3\sqrt{10}$$

Solutions: $3\sqrt{10}$, $-3\sqrt{10}$

Problems

1	$r^2 = 25$	**11**	$x^2 = 77$
2	$n^2 = -13$	**12**	$b^2 = 6$
3	$x^2 = 86$	**13**	$n^2 = -6$
4	$x^2 = 36$	**14**	$x^2 = 90$
5	$x^2 = 49$	**15**	$n^2 = 100$
6	$r^2 = -21$	**16**	$x^2 = 16$
7	$r^2 = 50$	**17**	$x^2 = 29$
8	$b^2 = 68$	**18**	$v^2 = 62$
9	$k^2 = 35$	**19**	$v^2 = 63$
10	$b^2 = -25$	**20**	$a^2 = -23$

RADICAL EQUATIONS

An equation containing a radical expression is called a radical equation. To solve a radical equations we have to apply the rules of exponents some basic algebraic rules.

$$\sqrt{2x + 1} = 1$$

$\sqrt{}$ *is the radical sign*

$2x + 1$ *is the radicand*

How to Solve

To find a solution to an equation which features a radical expression, we have:

1) to isolate our radical on only one side of the equation. After that, all we need to do is to square both sides of the equation

$$if\ x = y\ then\ x^2 = y^2$$

2) Square both sides to remove the radical

3) Once the Radical has been removed, solve the unknown

Example 1:

$$\sqrt{x} - 3 = 5$$

Add 3 to both sides to isolate the variable term on the left side of the equation

$$\sqrt{x} - 3 + 3 = 5 + 3$$

$$\sqrt{x} = 8$$

$$\left(\sqrt{x}\right)^2 = 8^2$$

$$x = 64 \text{ is the solution}$$

Square both sides to remove the radical, since $\left(\sqrt{x}\right)^2 = x$, make sure to square the 8 also, then simplify

After squaring a radical equation, it becomes common to get your solution to the squared equation that does not serve as a valid solution to the first equation.

There is a term for these equations which do not have the same answers as their original equations. This term is the extraneous solution.

These solutions should always be discarded everywhere that they turn up. This can be done only by checking the possible solutions of the original equation and comparing these to those of the squared equation. If these two do not match up, then it is our solution to the squared equation, or the extraneous solution, that should be discarded in these cases.

Example 2:

$$\sqrt{a-5} = -2$$

Square both sides to remove the term a-5 from the radical

$$\left(\sqrt{a-5}\right)^2 = (-2)^2$$

$$a - 5 = 4$$

$$a = 9$$

$$\sqrt{9-5} = -2$$

$$\sqrt{4} = -2$$

$$2 \neq -2$$

so, no solution

Example of extraneous Solution

$$\text{Solve } x + 4 = \sqrt{x + 10}$$

Square both sides to remove the term $x + 10$ from the radical

$$(x+4)^2 = \left(\sqrt{x+10}\right)^2$$

Now simplify and solve the equation

$$(x + 4)(x - 4) = x + 10$$

$$x^2 + 8x + 16 = x + 10$$

$$x^2 + 8x + 16 - x - 10 = 0$$

$$x^2 + 7x + 6 = 0$$

$$(x + 6)(x + 1) = 0$$

Set each factor equal to zero and solve for x

$$(x + 6) = 0 \quad or \quad (x + 1) = 0$$

$$x = -6 \quad or \quad x = -1$$

Now you check both solutions by substituting them into the original equation

$$-6 + 4 = \sqrt{-6 + 10} \qquad -1 + 4 = \sqrt{-1 + 10}$$

$$-2 = \sqrt{4} \qquad\qquad 3 = \sqrt{9}$$

$$-2 = 2 \; FALSE \qquad\quad 3 = 3 \; TRUE$$

Answer

$x = -1$ is the only solution

Example 5:

$$\sqrt{5x + 1} - 6 = 0$$

now isolate the x on the left

$$\sqrt{5x + 1} = 6$$

now square both isdes

$$\left(\sqrt{5x + 1}\right)^2 = 6^2$$

$$5x + 1 = 36$$

$$5x = 36 - 1$$

$$x = \frac{35}{5} = 7$$

Problems Solved

1) **Problem 1**

$$-2\sqrt{k+1} = 0$$

now divide both sides by 2, so

$$\frac{-2\sqrt{k+1}}{2} = \frac{0}{2}$$

$$-1\sqrt{k+1} = 0$$

now square both sides, so

$$k+1 = 0$$

$$k = -1$$

2) **Problem 11**

$$x = \sqrt{-14+9x}$$

now square both sides, so

$$x^2 = \left(\sqrt{-14+9x}\right)^2$$

$$x^2 = -14+9x$$

$$x^2 - 9x + 14 = 0$$

now find the roots of quadratic by factoring, so

$$(x-2)(x-7) \ \ so \ x = 2 \ \ and \ \ x = 7$$

Solution: 2, 7

Problems

1	$-2\sqrt{k+1} = 0$	**11**	$x = \sqrt{-14 + 9x}$
2	$-5 = -5 + \sqrt{x}$	**12**	$\sqrt{20 - k} = k$
3	$-5\sqrt{n} = -10$	**13**	$x = \sqrt{2x}$
4	$5 = \sqrt{n-5}$	**14**	$a = \sqrt{-32 + 12a}$
5	$4 = \sqrt{a}$	**15**	$\sqrt{2-n} = n$
6	$x = \sqrt{110 - x}$	**16**	$\sqrt{42 - v} = v$
7	$\sqrt{-1 - 2v} = v$	**17**	$\sqrt{n} = n$
8	$b = \sqrt{56 - b}$	**18**	$\sqrt{-40 + 13n} = n$
9	$\sqrt{30 - v} = v$	**19**	$\sqrt{30x - x} = x$
10	$\sqrt{72 - r} = r$	**20**	$x = \sqrt{30 + x}$

RATIONAL EQUATIONS

To solve you have to find the LCM that is the dividend of all denominators of the equation, then solve the equation

How to solve

Example 1:

The LCM is 8 because it is dividend of all denominators

$$\frac{x+2}{8} = \frac{6}{8}$$

now you should multiply each side of equation by LCM (8) to eliminate all the denominators, so

$$\cancel{8}\left(\frac{x+2}{\cancel{8}}\right) = \cancel{8}\left(\frac{6}{\cancel{8}}\right)$$

$$x + 2 = 6$$

$$x = 6 - 2$$

$$x = 4$$

Example 2:

$$\frac{7}{x+2} + \frac{5}{x-2} = \frac{10x-2}{x^2-4}$$

The LCM is Denominators: $x^2 - 4$

$$(x^2 - 4) = (x-2)(x+2)$$

Since $(x-2)$ *and* $(x+2)$ *are both factors of* $x^2 - 4$,

the LCM is $(x-2)(x+2)$, *so*

$$\frac{7(x-2)}{(x+2)(x-2)} + \frac{5(x+2)}{(x+2)(x-2)} = \frac{(10x-2)}{(x+2)(x-2)}$$

now multiply each side of equation by $(x+2)(x-2)$

to eliminate the denominators

$$\cancel{(x+2)}\cancel{(x-2)}\frac{7(x-2)}{\cancel{(x+2)}\cancel{(x-2)}} +$$

$$\cancel{(x+2)}\cancel{(x-2)}\frac{5(x+2)}{\cancel{(x+2)}\cancel{(x-2)}} =$$

$$\cancel{(x+2)}\cancel{(x-2)}\frac{(10x-2)}{\cancel{(x+2)}\cancel{(x-2)}}$$

Solve for x

$$7x - 14 + 5x + 10 = 10x - 2$$

$$7x + 5x - 10x = 14 - 10 - 2$$

$$2x = 2$$

$$x = \frac{2}{2} = 1$$

Problems Solved

1) **Problem 1**

$$\frac{1}{4} + \frac{3}{4b} = \frac{b-2}{2b}$$

$$LCM = 4b$$

$$\frac{b+3}{4b} = \frac{2(b-2)}{4b}$$

$$\frac{b+3}{4b} = \frac{2b-4}{4b}$$

$$4b\left(\frac{b+3}{4b}\right) = 4b\left(\frac{2b-4}{4b}\right)$$

$$b+3 = 2b-4$$

$$b-2b = -4-3$$

$$-b = -7$$

Solution $b = 7$

2) Problem 3

$$\frac{1}{4m} + \frac{1}{2} = \frac{1}{4}$$

$$\text{LCM} = 4m$$

$$\frac{1 + 2m}{4m} = \frac{m}{4m}$$

$$4m\left(\frac{1 + 2m}{4m}\right) = 4m\left(\frac{m}{4m}\right)$$

$$1 + 2m = m$$

$$2m - m = -1$$

$$m = -1$$

3) Problem 2

$$\frac{4}{3v^2} - \frac{1}{v} = \frac{v - 4}{v^2}$$

$$LCM = 3v^2$$

$$\frac{4 - 3v}{3v^2} = \frac{3(v - 4)}{3v^2}$$

$$\frac{4 - 3v}{3v^2} = \frac{3v - 12}{3v^2}$$

$$(3v^2)\left(\frac{4 - 3v}{3v^2}\right) = (3v^2)\frac{3v - 12}{3v^2}$$

$$4 - 3v = 3v - 12$$

$$-3v - 3v = -12 - 4$$

$$-6v = -16$$

$$6v = 16$$

$$v = \frac{16}{6} = \frac{8}{3}$$

$$\textit{Solution } v = \frac{8}{3}$$

Problems

1	$\dfrac{1}{4} + \dfrac{3}{4b} = \dfrac{b-2}{2b}$
2	$\dfrac{4}{3v^2} - \dfrac{1}{v} = \dfrac{v-4}{v^2}$
3	$\dfrac{1}{4m} + \dfrac{1}{2} = \dfrac{1}{4}$
4	$\dfrac{2}{x^2} + \dfrac{1}{2x} = \dfrac{1}{x^2}$
5	$\dfrac{n+5}{n^2} - \dfrac{1}{n^2} = \dfrac{1}{2n}$
6	$\dfrac{1}{x} - \dfrac{3}{x^2} = \dfrac{5x+5}{x^2}$
7	$\dfrac{6}{x^2} - \dfrac{2x-6}{x^2} = \dfrac{1}{x}$
8	$\dfrac{2}{x^2} = \dfrac{1}{6x^2} + \dfrac{1}{2x}$
9	$\dfrac{1}{2r} = \dfrac{3}{2r^2} + \dfrac{r-5}{r^2}$
10	$\dfrac{n+4}{6n} - \dfrac{1}{2n} = \dfrac{1}{3n}$

11	$$\frac{x^2 + x - 30}{5x} = \frac{x + 5}{5} + \frac{1}{5x}$$
12	$$\frac{5}{n^3} = \frac{n + 2}{n^2} - \frac{n - 3}{n^3}$$
13	$$\frac{n}{4} - \frac{1}{2n} = \frac{n^2 - 3n - 18}{4n}$$
14	$$\frac{b + 5}{2} + \frac{4}{b} = \frac{b + 3}{2}$$
15	$$\frac{x^2 - 3x - 10}{2x^2} = \frac{2}{x^2} + \frac{1}{x}$$
16	$$\frac{1}{6n^2 - 16n + 10} = \frac{1}{3n^2 - 8n + 5} + \frac{1}{2n - 2}$$
17	$$\frac{1}{n^2 + 3n} + \frac{n + 7}{n^2 + 3n} = \frac{5n + 20}{3n^2 + 9n}$$
18	$$\frac{1}{n - 4} - \frac{1}{n^2 - 4n} = \frac{2}{n^2 - 4n}$$
19	$$\frac{4}{3r^2 + 24r + 45} = \frac{1}{r^2 + 8r + 15} + \frac{1}{3r + 9}$$
20	$$\frac{1}{n} = \frac{7}{n} - \frac{3n + 3}{n^2 - 2n}$$

LINEAR EQUATIONS

A linear equation can be written in the following form:

$$ax + b = 0$$

a and b are real numbers and x is the variable

Properties:

1) *if $a = b$ then $a + c = b + c$ for any c*
2) *if $a = b$ then $a - c = b - c$ for any c*
3) *if $a = b$ then $ac = bc$ for any c*
4) *if $a = b$ then $a/c = b/c$ for any c*

How to Solve

1) If the equation contains fractions use the LCM to clear the fractions. Multiply the LCM by both sides of the equation
2) Simplify both sides
3) Use the first two properties to get all terms with the variable on one side of the equation and all constants on the other side.
4) If the coefficient of the variable is not a one, use the third or fourth property to make the coefficient a 1.

Example 1:

$$4x - 7(2 - x) = 3x + 2$$

$$4x - 14 + 7x = 3x + 2$$

$$11x - 14 = 3x + 2$$

$$8x = 16$$

$$x = \frac{16}{8}$$

$$x = 2$$

Example 2:

$$\frac{5x}{3x - 3} + \frac{6}{x + 2} = \frac{5}{3}$$

LCM is $3(x - 1)(x + 2)$

$$3(x - 1)(x + 2)\left(\frac{5x}{3(x - 1)} + \frac{6}{x + 2}\right) =$$

$$\left(\frac{5}{3}\right)[3(x - 1)(x + 2)]$$

$$5x(x + 2) + 3(x - 1)(6) = 5(x - 1)(x + 2)$$

$$5x^2 + 10x + 18(x - 1) = 5(x^2 + x - 2)$$

$$5x^2 + 10x + 18x - 18 = 5x^2 + 5x - 10$$

$$5x^2 + 10x + 18x - 18 - 5x^2 = 5x - 10$$

$$28x - 18 = 5x - 10$$

$$28x = 8$$

$$x = \frac{8}{23}$$

Example 3:

$$\frac{5x + 8}{7} + \frac{7x - 3}{5} = \frac{6x + 39}{7}$$

$$\frac{5x + 8}{7} + \frac{7x - 3}{5} - \frac{6x + 39}{7} = 0$$

$$\frac{25x + 40 + 49x - 21 - (30x + 195)}{35} = 0$$

$$\frac{25x + 40 + 49x - 21 - 30x - 195}{35} = 0$$

$$\frac{x(25 + 49 - 30) + 40 - 21 - 195}{35} = 0$$

$$44x - 176 = (0 \cdot 35)$$

now we move the denominator 35 on the

right side, so we multiply it by 0

now we isolate the x on the left side, so

$$44x = 176$$

$$x = \frac{176}{44}$$

$$x = 4$$

Problems

1) $3(x + 5) = 2(-6 - x) - 2x$

2) $-(-3 + x) + 4 = 2(x + 3) - 6x + 7$

3)

$$-\frac{3}{4} - \frac{1}{2}x + \frac{3}{4} = \frac{5}{6} - \frac{3}{4}$$

4)
$$\frac{m - 2}{3} + 1 = \frac{2m}{7}$$

Problems with variable in the denominators

5)
$$\frac{2x}{x + 3} = \frac{3}{x - 10} + 2$$

6)
$$\frac{2}{x + 2} = \frac{-x}{x^2 + 5x + 6}$$

EQUATION SYSTEMS

A linear system of two equations with 2 variables is any system written in the following form:

$$ax + by = p$$

$$cx + dy = q$$

Other examples of systems with numbers are:

$$3x - y = 7$$

$$2x + 3y = 1$$

There are two methods for resolving the Equation Systems:

1) Substitution Method
2) Elimination Method

Substitution Method

The first method is called the method of substitution. In this method we have to solve first one of the equations for one of the variables and then we have to substitute this into the other equation.

How to Solve

Example 1

$$3x - y = -2$$
$$x - 2y = 6$$

you can start by choosing one of the two equations; in this case it's easier to start with the second equation

$$x = 6 + 2y$$

Step 1

now replace the x of the first equation with $6 + 2y$

$$3(6 + 2y) - y = -2$$

Step 2

$$18 + 6y - y = -2$$
$$5y = -2 - 18$$
$$5y = -20$$
$$y = \frac{-20}{5} = -4$$

Step 3

now replace the y of the second equation

with the value $- 4$

$$x - 2(-4) = 6$$
$$x + 8 = 6$$
$$x = 6 - 8 = -2$$

Solution $(-2, -4)$

Problems Solved

1) Problem 16

$$3x - 2y = 8$$

$$x + y = 1$$

Step 1

$$x = 1 - y$$

Step 2

$$3(1 - y) - 2y = 8$$

$$3 - 3y - 2y = 8$$

$$3 - 5y = 8$$

$$-5y = 8 - 3$$

$$-5y = 5$$

now we change the sign of y by changing all the signs

$$5y = -5$$

$$y = -\frac{5}{5} = -1$$

Step 3

$$x + (-1) = 1$$

$$x - 1 = 1$$

$$x = 2$$

Solution $(2, -1)$

2)   Problem 10

$$3x + 4y = 8$$
$$x - 2y = 6$$

Step 1

$$x = 6 + 2y$$

Step 2

$$3(6 + 2y) + 4y = 8$$
$$18 + 6y + 4y = 8$$
$$10y = 8 - 18$$
$$10y = -10$$
$$y = \frac{-10}{10} = -1$$

Step 3

$$x - 2(-1) = 6$$
$$x + 2 = 6$$
$$x = 4$$

Solution $(4, -1)$

3) Problem 3

$$7x + 3y = 12$$
$$x - 3y = 12$$

$$x = 12 + 3y$$

$$7(12 + 3y) + 3y = 12$$
$$84 + 21y + 3y = 12$$
$$24y = 12 - 84$$
$$24y = -72$$
$$y = \frac{-72}{24} = -3$$

$$x - 3(-3) = 12$$
$$x + 9 = 12$$
$$x = 3$$

Solution $(3, -3)$

Problems

1	$3x - y = -2$ $x - 2y = 6$	**11**	$2x - y = -3$ $5x + y = -4$
2	$3x - y = 4$ $5x + y = 4$	**12**	$x + y = -4$ $4x - y = -1$
3	$7x + 3y = 12$ $x - 3y = 12$	**13**	$x = -3$ $x - 3y = 3$
4	$x - 2y = 4$ $2x + y = 3$	**14**	$6x + y = -3$ $x + y = 2$
5	$5x - y = 1$ $x - y = -3$	**15**	$x + y = 4$ $x + y = -1$
6	$x - y = -3$ $5x - y = 1$	**16**	$3x - 2y = 8$ $x + y = 1$
7	$x - y = -3$ $7x - y = 3$	**17**	$x - 2y = 6$ $5x - 2y = -2$
8	$x + 2y = -6$ $5x - 4y = -16$	**18**	$4x + y = 3$ $4x + y = -2$
9	$x + 2y = -4$ $2x + y = 4$	**19**	$5x + y = 1$ $x + y = -3$
10	$3x + 4y = 8$ $x - 2y = 6$	**20**	$3x - 2y = -8$ $3x - 2y = 2$

Elimination Method

Steps of the Method:

1) to multiply one or both of the equations by appropriate numbers, in this case one of the variables have the same coefficient but with opposite signs.

2) Add the two equations together

3) Since one of the variables had the same coefficient with opposite signs, it will be eliminated when we add the two equations.

4) Solve the single equation for one of the variables.

5) Substitute the previous equation solution value into one of the original equations.

How to Solve

Example 1:

$$x + y = 2$$
$$x - y = 14$$

As you see, if you add the variable y they will be 0, so they are eliminated. In this case you can add immediately the two equations, so

Step 1   adding equations, so:

$$x + y = 2 +$$
$$x - y = 14$$
$$\overline{2x = 16}$$

Step 2   Solve for x

$$2x = 16$$
$$x = \frac{16}{2} = 8$$

Step 3   Solve for y by replacing the value of x (8) with the x in the first equation, so:

$$x + y = 2$$
$$8 + y = 2$$

$$y = 2 - 8 = -6$$

Solution   (8, -6)

Problems Solved

1) **Problem 2**

$$5x + y = 3$$

$$x + y = -1$$

As you see, if you add the two equations, any variable will be eliminated. In this case you multiply the second equation by the coefficient of x in the first equation, so

$$5(x + y) = -1(5)$$

$$5x + 5y = -5$$

Step 1   To eliminate x you must subtract the two equations,

so

$$5x + y = 3$$

$$-(5x + 5y = -5)$$

$$\overline{-4y = 8}$$

$$4y = -8$$

$$y = \frac{-8}{4} = -2$$

Step 2   Solve for x by replacing the value of y (-2)

in the x of the first equation, so

$$5x + (-2) = 3$$

$$5x - 2 = 3$$

$$5x = 5$$

$$x = \frac{5}{5} = 1$$

Solution $(1, -2)$

Problem 18

$$6x - y = 2$$
$$x - y = -3$$

As you see, if you add the two equations, any variable will be eliminated. In this case you multiply the second equation by the coefficient of x in the first equation, so

$$6(x - y) = 6(-3)$$
$$6x - 6y = -18$$

Step 1   To eliminate x you must subtract

the two equations,

so

$$6x - y = 2$$
$$-(6x - 6y = -18)$$

$$5y = 20$$
$$y = \frac{20}{5} = 4$$

Step 2   Solve for x by replacing the value of y (4)in the x

of the first equation, so

$$6x - 4 = 2$$
$$6x = 6$$
$$x = \frac{6}{6} = 1$$

Solution $(1, 4)$

2) Problem 1

$$x + y = -4$$
$$5x - y = -2$$

As you see, if you add the variable y they will be 0, so they are eliminated. In this case you can add immediately the two equations, so

$$x + y = -4$$
$$5x - y = -2$$

$$6x = -6$$
$$x = \frac{-6}{6} = -1$$

Now you can solve for y by replacing the value of x (-1) in the x of the second equation, so

$$5(-1) - y = -2$$
$$-5 - y = -2$$
$$-y = 5 - 2$$
$$-y = 3$$
$$y = -3$$

Solutions (-1, -3)

Problems

1	$x + y = -4$ $5x - y = -2$	**11**	$x - y = 4$ $6x + y = 3$
2	$5x + y = 3$ $x + y = -1$	**12**	$3x + 2y = 6$ $3x + 2y = 2$
3	$y = 2$ $3x - 4y = 4$	**13**	$3x + 2y = 6$ $3x + 2y = 8$
4	$2x + 3y = -3$ $2x - 3y = -9$	**14**	$x + y = -1$ $x - 4y = -16$
5	$2x - y = -4$ $x - 2y = 4$	**15**	$x - 4y = 16$ $x - 4y = 8$
6	$x - 2y = 2$ $2x - y = -2$	**16**	$x + 2y = -4$ $3x + y = 3$
7	$5x - 3y = 12$ $x + y = 4$	**17**	$x + 3y = 9$ $2x + y = -2$
8	$5x - 2y = -4$ $x - 2y = 4$	**18**	$6x - y = 2$ $x - y = -3$
9	$x - 3y = 6$ $7x - 3y = -12$	**19**	$x + 4y = 8$ $x + y = -1$
10	$x + 2y = 2$ $5x + 2y = -6$	**20**	$x - 2y = -8$ $3x + 4y = -4$

Graphing Method

When systems of linear equations are graphed, the two equations are expressed by two individual lines, with the point at which the two meets up representing the solutions to the system.

We will now start by solving for our linear equation system that follows:

$$y = 2x + 4$$
$$y = 3x + 2$$

When graphed this system would look something like this:

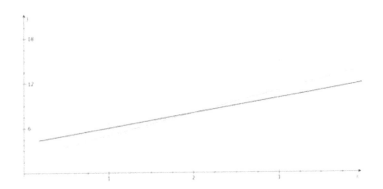

As you can see, the coordinate pair where these two lines are intersecting is about at x = 2, and y = 8. With this being said, our solution to this linear equation system would then be (2, 8).

A system of equations that are linear is multiple linear equations within one problem.

For instance:

$$y = 0.5x \quad and \quad y = x - 2$$

In order to determine our solution of this equation system, we would have to find an ordered pair that would solve for both equations. After we have ascertained the solution to both equations in question, our answer will at the point at which both of the lines that we have to graph intersect.

When graphed

$$y = 0.5x + 2 \quad and \quad y = x - 2$$

look like this:

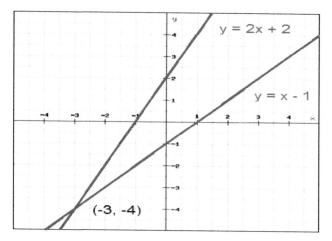

As the graph shows, the point at which these lines intersect is (-3, -4), which would then be the solution to this system of equations.

Example 2:

Let's now solve for another system:

$$y = -x + 7$$

$$y = 2x + 1$$

First, we need to find the y-intercepts of both lines, which are expressed by b in the standard equation y = mx + b. As you can see, these are 7 and 1. Next, we need to determine the slopes of both lines, which are expressed by m in the equation

$$y = mx + b$$

These are -1 and 2 in this case. We would now graph both lines, leaving us with this:

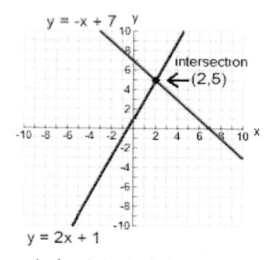

Our intersection here is (2, 5), which would be the solution to this system of equations.

Learning by what means one solves equations that are linear within a coordinate plane is very useful in the study of geometry and, later, even some statistical analysis.

Solving for Equation Systems with Graphing

A system of equations that are linear is multiple linear equations within one problem. For instance, y=0.5x+2 and y=x-2. In order to determine our solution of this equation system, we would have to find an ordered pair that would solve for both equations. After we have ascertained the solution to both equations in question, our answer will at the point at which both of the lines that we have to graph intersect.

When graphed, y=0.5x+2 and y=x-2 look like this:

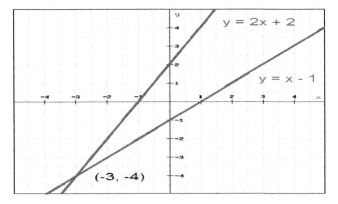

As the graph shows, the point at which these lines intersect is (-3, -4), which would then be the solution to this system of equations.

INEQUALITIES

Inequalities type $<$ and $\leq$

For example :

$$|p| \leq 4$$

With this notation, no matter what p is it, but we know that it must have a distance no more than 4 from the origin.

In other words

$$-4 \leq p \leq 4$$

Hereafter the general formula

$$if \ |p| \leq b, \quad b > 0 \quad then -b \leq p \leq b$$

$$if \ |p| < b, \quad b > 0 \quad then \ -b < p < b$$

$$if \ |p| \geq b, \quad b > 0 \quad then -p \leq -b \ or \ p \geq b$$

$$if \ |p| > b, \quad b > 0 \quad then \ p < -b \ or \ p > b$$

How to Solve

$$|2x - 3| > 7$$

$$2x - 3 < -7 \quad or \quad 2x - 3 > 7$$

$$2x < -4 \quad or \quad 2x > 10$$

$$x < -2 \quad or \quad x > 5$$

The interval notation are $(-\infty, -2)$ or $(5, \infty)$

$$|6t + 10| \geq 3$$

$$6t + 10 \leq -3 \quad or \quad 6t + 10 \geq 3$$

$$6t \leq -13 \quad or \quad 6t \geq -7$$

$$t \leq -\frac{13}{6} \quad or \quad t \geq -\frac{7}{6}$$

The interval notation are $\left(-\infty, \frac{13}{6}\right]$ or $\left[-\frac{7}{6}, \infty\right)$

Inequality	Graph	Interval Notation
$a \leq x \leq b$		$[a, b]$
$a < x < b$		(a, b)
$a \leq x < b$		$[a, b)$
$a < x \leq b$		$(a, b]$
$x > a$		(a, ∞)
$x \geq a$		$[a, \infty)$
$x < b$		$(-\infty, b)$
$x \leq b$		$(-\infty, b]$

- a bracket, "[" or "]", means that we include the endpoint
- a parenthesis, "(" or ")", means we don't include the endpoint

Rules for Solutions

1) If a < b then a + c < b + c and a − c < b − c for any number c

2) If a < b and c > 0 then ac < bc and $\dfrac{a}{c} < \dfrac{b}{c}$

3) If a < b and c < 0 then ac > bc and $\dfrac{a}{c} > \dfrac{b}{c}$

Graphing inequality on line number

1) use a filled-on circle to include a number

2) use open circle to exclude it

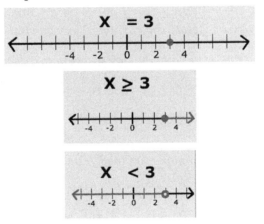

Absolute Value

How to solve

Example 1:

$$|2x - 4| < 10$$

$$if \ |p| < b, \qquad b > 0 \qquad then \quad -b < p < b$$

We can write as follows

$$-10 < 2x - 4 < 10$$

Now we isolate the variable 2x by moving

−4 on the left and right side by changing its sign

$$-10 + 4 < 2x < 10 + 4$$

$$-6 < 2x < 14$$

Now we divide the number on the left and on the right side
by the coefficient of x, (2)

$$\frac{-6}{2} < 2 < \frac{14}{2}$$

$$-3 < x < 7$$

The interval notation is $(-3, 7)$

Example 2:

$$|9m + 2| \leq 1$$

$$9m + 2 \leq 1$$

$$if \ |p| \leq b, \quad b > 0 \quad then \ -b \leq p \leq b$$

$$-1 - 2 \leq 9m + 2 \leq 1 - 2$$

$$-3 \leq 9m \leq -1$$

$$\frac{-3}{9} \leq 9 \leq \frac{-1}{9}$$

The interval notation is $\left[-\dfrac{1}{3}, -\dfrac{1}{9}\right]$

Problems Solved with graphics

1) **Problem 1**

$$|v - 7| > 5$$

$$if\ |p| > b, \qquad b > 0 \qquad then\ p < -b\ or\ p > b$$

$$-5 > v - 7 > 5$$

$$-5 + 7 > v > 5 + 7$$

$$2 > v > 12$$

The interval notation is $(-\infty, 2)\ or\ (12, \infty)$

2) **Problem 2**

$$|a - 4| \geq 9$$

$$a - 4 \geq 9$$

$$a \geq 9 + 4$$

$$a \geq 13$$

$$OR$$

$$a - 4 \leq -9$$

$$a \leq -9 + 4$$

$$a \leq -5$$

The interval notation is $(-\infty, -5]\quad or\quad [13, \infty)$

234

3) Problem 6

$$\left|-\frac{3}{4}p\right| \le \frac{15}{28}$$

$$-\frac{3}{4}p \le \frac{15}{28}$$

we can change the signs on both sides, so

$$\frac{3}{4}p \le -\frac{15}{28}$$

$$p \le -\frac{15}{28} \cdot \frac{4}{3}$$

simply with criss cross, so

$$p \le -\frac{5}{7}$$

OR

$$\frac{3}{4}p \ge -\frac{15}{28}$$

$$p \ge -\frac{15}{28} \cdot \frac{4}{3}$$

$$p \ge -\frac{5}{7}$$

the interval notation is

$$-\frac{5}{7} \le p \le \frac{5}{7}$$

$$\left[-\frac{5}{7}, \frac{5}{7}\right]$$

Problems

Solve each inequality and graph its solution

1	$|v - 7| > 5$
2	$|a - 4| \geq 9$
3	$\left|\dfrac{b}{5}\right| \geq 5$
4	$|-6 + r| < 5$
5	$|v - 5| \geq 7$
6	$\left|-\dfrac{3}{4}p\right| \leq \dfrac{15}{28}$
7	$\left|p - \dfrac{7}{3}\right| \leq \dfrac{7}{12}$
8	$\left|\dfrac{5}{3}n\right| \leq \dfrac{5}{7}$
9	$\left|x + \dfrac{39}{8}\right| \geq \dfrac{59}{40}$
10	$\left|n + \dfrac{5}{8}\right| > \dfrac{33}{8}$

Compound

A compound inequality is an equality which contains at least two inequalities separated by either "and" or "or".

How to solve

Example 1:

$$y - 3 > 5 \quad or \quad y + 3 < -2$$
$$y - 3 + 3 > 5 + 3 \quad or \quad y + 3 - 3 < -2 - 3$$
$$y > 8 \quad or \quad y < -5$$

Solution is $y > 8 \quad or \quad y < -5$

Now we can represent both solutions on the number line

Problems Solved

1)   Problem 10

$$4x < 16 \ \ and \ \ 2 + x > -8$$

$$4x < 16 \ \ and \ \ 2 + x > -8$$

$$x < \frac{16}{4} \ \ and \ \ x > -8 - 2$$

$$x < 4 \ \ and \ \ x > -10$$

so

$$-10 < x < 4$$

2)   Problem 6

$$10n \le 30 \ \ and \ \ \frac{n}{10} > 0$$

$$n \le \frac{30}{10} \ \ and \ \ n > \frac{0}{10}$$

$$n \le 3 \ \ and \ \ n > 0$$

so

$$0 < n \le 3$$

Problems

1	$\dfrac{v}{7} > 1 \ \ or \ \ v - 10 \leq -13$
2	$\dfrac{x}{6} < 0 \ or \ x + 9 \geq 10$
3	$8n < 40 \ and \ n + 4 \geq 4$
4	$-1 \leq x - 7 \leq 2$
5	$n - 6 > -2 \ and \ 3n < 27$
6	$10n \leq 30 \ and \ \dfrac{n}{10} > 0$
7	$k + 3 > 2 \ and \ \dfrac{k}{9} < 1$
8	$\dfrac{a}{2} \geq -2 \ and \ a - 5 \leq -5$
9	$\dfrac{x}{3} > 0 \ or \ \dfrac{x}{5} < -2$
10	$4x < 16 \ and \ 2 + x > -8$

11	$3m \geq \dfrac{17}{2}$ or $\dfrac{5}{3}m < -\dfrac{16}{3}$
12	$n + \dfrac{7}{4} \geq -\dfrac{3}{4}$ and $\dfrac{7}{24}n < -\dfrac{35}{192}$
13	$-3x \leq -\dfrac{2}{3}$ or $x + \dfrac{11}{7} \leq -\dfrac{3}{7}$
14	$\dfrac{72}{5} > \dfrac{9}{5}m > \dfrac{9}{40}$
15	$a + \dfrac{9}{2} \leq \dfrac{29}{6}$ or $a + 8 > \dfrac{119}{10}$
16	$n - \dfrac{5}{4} \leq \dfrac{165}{28}$ and $\dfrac{2}{3}n \geq \dfrac{14}{3}$
17	$-\dfrac{1}{2} \leq \dfrac{1}{2}x < 3$
18	$\dfrac{7}{11}m \geq -\dfrac{14}{11}$ and $-\dfrac{7}{5}m > -\dfrac{48}{5}$
19	$v + \dfrac{9}{8} > -\dfrac{129}{56}$ and $v + \dfrac{29}{6} < \dfrac{121}{12}$
20	$x - \dfrac{1}{2} \geq \dfrac{7}{6}$ or $x - 2 \leq -\dfrac{3}{2}$

Graphing

How to solve

Solve the inequality

$$x + 3 < 7$$

$$x + 3 - 3 < 7 - 3$$

$$x < 4$$

A different way to state this answer is stating that x equals some value that is less than 4.

The picture below illustrates what we have done here on a number line:

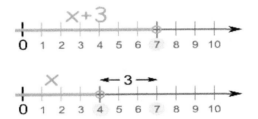

Quick method

Case 1) variable x is positive

$$x + 3 < 7$$

You move the number 3 to the right side of inequality symbols by changing the sign, so

$$x < 7 - 3$$

$$x < 4$$

Case 2) variable x is negative

$$-x \leq 2$$

You can move the x on the right side of the inequality symbol by changing its sign and move the number on the left side by changing its sign, so

$$-2 \leq x$$

that we can read as x is higher or equal to − 2

Problems Solved

1) **Problem 5**

$$-n < 0$$

$$n > 0$$

2) **Problem 2**

$$-n \geq -3$$

$$3 \geq n$$

the same thing if we write $n \leq 3$

3) **Problem 12**

$$-n < \frac{1}{2}$$

$$\frac{1}{2} = 0.5$$

$$-n < 0.5$$

$$-0.5 < n$$

that we can read as n greater than -0.5

The following are examples with **mixed numbers**, to be converted into **improper fractions**

Example 4: this is the problem 13 of the next chapter

$$-x \leq -2\frac{1}{2}$$

$-2\frac{1}{2}$ *is a mixed number, so we must convert it*

*in an **improper fraction** in this way:*

$$-2\frac{1}{2} = -\frac{(2 \cdot 2) + 1}{2} = -\frac{5}{2}$$

$$-\frac{5}{2} = -2.5 \text{ so}$$

$$-x \leq -2.5, \text{ that is}$$

$$x \geq 2.5$$

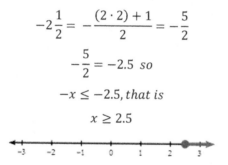

Example 5: this is the problem 19 of the next chapter

$$k \leq -1\frac{1}{2}$$

$$k \leq -\frac{(2 \cdot 1) + 1}{2} = -\frac{3}{2} = -1.5$$

$$k \leq -1.5$$

Problems

Draw a graph for each inequality

1	$b \geq -2$	11	$n \leq -1$
2	$-n \geq -3$	12	$-n < \dfrac{1}{2}$
3	$-x \leq 2$	13	$-x \leq -2\dfrac{1}{2}$
4	$m \geq 6$	14	$-m \leq \dfrac{1}{2}$
5	$-n < 0$	15	$-x < 2\dfrac{1}{2}$
6	$m > -6$	16	$-m \leq 0$
7	$x > 5$	17	$b > \dfrac{1}{2}$
8	$x \leq 4$	18	$m \leq \dfrac{1}{2}$
9	$x < 6$	19	$k \leq -1\dfrac{1}{2}$
10	$-x > 2$	20	$-x \geq 2$

One Step Inequalities

Let's now start with subtracting and adding inequalities. This is usually done by subtracting or adding the number in question on both sides of the equation, or of the inequality marker. This will help to isolate the variable(s) that we are looking for.

How to Solve

Example 1: $x + 2 < 7$

To solve for x within our inequality we subtract both sides by 2

$$x + 2 - 2 < 7 - 2$$
$$x < 5$$

Quick method

Instead of subtract on both sides the number 2, I prefer to change the sign when I move on the other side of inequality symbols, so

$$x + 2 < 7$$
$$x < 7 - 2$$
$$x < 5$$

Problems Solved

1) **Problem 4**

$$0 < p + 14$$
$$0 - 14 < p$$
$$p > -14$$

2) **Problem 1**

$$\frac{x}{19} \geq -6$$

*when you move a denominator in the other side,
it becomes a numerator to multiply*

$$x \geq -6 \cdot 19$$
$$x \geq -114$$

Problems

1	$\dfrac{x}{19} \geq -6$	**11**	$-\dfrac{99}{10} < p - 13$
2	$42 < -7n$	**12**	$\dfrac{10915}{102} \geq \dfrac{177}{17}r$
3	$\dfrac{x}{3} > -20$	**13**	$\dfrac{9}{5}p \leq 3$
4	$0 < p + 14$	**14**	$-\dfrac{27}{16} < 2 + v$
5	$26 > n + 11$	**15**	$\dfrac{4v}{7} \geq -\dfrac{6}{7}$
6	$a - 9 \geq -20$	**16**	$b - \dfrac{7}{3} < -\dfrac{13}{3}$
7	$\dfrac{p}{12} \geq \dfrac{5}{6}$	**17**	$\dfrac{7}{3}a < -\dfrac{49}{12}$
8	$r + 15 < 24$	**18**	$\dfrac{33}{10} > -1 + x$
9	$-15 + x < -24$	**19**	$\dfrac{196}{15} \geq r - \dfrac{29}{15}$
10	$0 > 20 + x$	**20**	$x - \dfrac{11}{6} < -\dfrac{17}{6}$

Two Step Inequalities

How to solve

Example 1:

$$3x + 5 > 17$$
$$3x + 5 - 5 > 17 - 5$$
$$3 > 12$$
$$x > 4$$

Solution: x > 4 so any value greater than 4 will make the inequality correct

Quick method:

Instead of adding on both sides the number -5 I prefer to move the number 5 on the other side by changing it sign, so

$$3x > 17 - 5$$
$$3x > 12$$
$$x > \frac{12}{3}$$
$$x > 4$$

Problems Solved

1) Problem 9

$$10 - 4m \leq -18$$

$$-4m \leq -18 - 10$$

$$-4m \leq -28$$

when you change signs on both sides, you must reverse the inequality symbol

$$4m \geq 28$$

$$m \geq \frac{28}{4}$$

$$m \geq 7$$

2) Problem 1

$$\frac{x}{2} + 7 \leq 12$$

$$\frac{x}{2} \leq 12 - 7$$

$$\frac{x}{2} \leq 5$$

$$x \leq 5 \cdot 2$$

$$x \leq 10$$

250

3) Problem 6

$$1 > \frac{n+6}{9}$$

when you move the denominator on the other side it
becomes a numerator to be multiplied

$$1 \cdot 9 > n + 6$$

$$9 > n + 6$$

$$9 - 6 > n$$

$$3 > n$$

that is

$$n < 3$$

4) Problem 12

$$-\frac{71}{4} > 5 + 1\frac{3}{10}v$$

$1\frac{3}{10}$ *is a mixed number, so we convert it*
into an improper fraction, so

$$1\frac{3}{10} = \frac{(10 \cdot 1) + 3}{10} = \frac{13}{10}$$

$$-\frac{71}{4} > 5 + \frac{13}{10}v$$

$$-\frac{71}{4} - 5 > \frac{13}{10}v$$

$$\frac{-71-20}{4} > \frac{13}{10}v$$

$$-\frac{91}{4} > \frac{13}{10}v$$

that is the same if we write

$$\frac{13}{10}v < -\frac{91}{4}$$

$$v < -\frac{91}{4} \cdot \frac{10}{13}$$

now we can simplify with criss − cross, so

$$v < -\frac{35}{2}$$

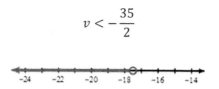

Problems

1	$\dfrac{x}{2} + 7 \leq 12$	**11**	$4\dfrac{7}{10}x - \dfrac{3}{2} < \dfrac{2753}{170}$
2	$5 + 6x < -115$	**12**	$-\dfrac{71}{4} > 5 + 1\dfrac{3}{10}v$
3	$66 > 10 - 7n$	**13**	$\dfrac{118}{9} > -\dfrac{1}{9} + 1\dfrac{1}{3}a$
4	$\dfrac{-6 + n}{5} > -3$	**14**	$-1\dfrac{4}{5}k + 5\dfrac{7}{10} \geq \dfrac{213}{50}$
5	$-4(-1 + k) \leq 28$	**15**	$\dfrac{357}{16} > 3\dfrac{1}{2}\left(x - \dfrac{1}{8}\right)$
6	$1 > \dfrac{n + 6}{9}$	**16**	$\dfrac{91}{64} \geq \dfrac{1}{4}x - \dfrac{3}{4}$
7	$\dfrac{v}{4} + 9 \leq 6$	**17**	$\dfrac{14}{9} - \dfrac{17}{10}n \geq -\dfrac{17777}{1260}$
8	$6n - 8 \geq -74$	**18**	$-\dfrac{1429}{200} > -\dfrac{8}{5}m + \dfrac{3}{8}$
9	$10 - 4m \leq -18$	**19**	$\dfrac{7}{20} \leq \dfrac{3}{2}x + \dfrac{8}{5}$
10	$\dfrac{r + 7}{8} \leq 1$	**20**	$-\dfrac{549}{104} > 4\dfrac{1}{2}\left(n - \dfrac{1}{4}\right)$

Multi Step Inequalities

How to solve

1) **Problem 9**

$$-6(7n + 8) - 6n < -432$$

$$-42n - 48 - 6n < -432$$

$$-48n < -432 + 48$$

$$-48n < -384$$

when both sides are negative you chage them in positive

and reverse the inequality symbol

$$48n > 384$$

$$n > \frac{384}{48}$$

$$n > 8$$

2) Problem 10

$$231 < 7(1 + 4b)$$
$$231 < 7 + 28b$$
$$231 - 7 < 28b$$
$$224 < 28b$$
$$\frac{224}{28} < b$$

when you move a numerator in the other side it becomes divisor

$$8 < b$$

that is the same as

$$b > 8$$

3) Problem 1

$$148 \leq 5(1 + 5x) - 7$$
$$148 \leq 5 + 25x - 7$$
$$148 \leq 25x - 2$$
$$148 + 2 \leq 25x$$
$$150 \leq 25x$$

when you move a factor to the other side, it becomes divisor

$$\frac{150}{25} \leq x$$
$$6 \leq x$$
$$x \geq 6$$

4)   Problem 12

$$-\frac{864}{7} > \frac{32}{7}\left(\frac{7}{2}r + 1\right)$$

$$-\frac{864}{7} > \frac{224}{14}r + \frac{32}{7}$$

$$-\frac{864}{7} - \frac{32}{7} > \frac{224}{14}r$$

$$-\frac{896}{7} > \frac{224}{14}r$$

$$-\frac{896}{7} \cdot \frac{14}{224} > r$$

now we can simplify with criss cross

$$-\frac{4}{1} \cdot \frac{2}{1} > r$$

$$-8 > r$$

that is

$$r > -8$$

Problems

1	$148 \leq 5(1 + 5x) - 7$
2	$-6(-4x + 6) \geq 156$
3	$5(5n - 7) - 3n \leq -189$
4	$-173 > -5(7 - 6n) - 7n$
5	$7(n + 7) > 91$
6	$296 \geq 8(-6x + 7)$
7	$-1 - 7(n + 6) < -99$
8	$6 + 7(r - 7) \geq -92$
9	$-6(7n + 8) - 6n < -432$
10	$231 < 7(1 + 4b)$

11	$7\left(\dfrac{25}{7}n + \dfrac{7}{5}\right) \le -\dfrac{2581}{30}$
12	$-\dfrac{864}{7} > \dfrac{32}{7}\left(\dfrac{7}{2}r + 1\right)$
13	$-\dfrac{22}{3}\left(2n + \dfrac{3}{7}\right) \ge \dfrac{594}{7}$
14	$-8\left(\dfrac{5}{8}x + 7\right) - \dfrac{5}{3}x \le -\dfrac{511}{6}$
15	$-\dfrac{13}{4}\left(5x - \dfrac{23}{8}\right) < \dfrac{3419}{32}$
16	$-\dfrac{10785}{112} \ge -\dfrac{15}{4}\left(\dfrac{15}{4}r - \dfrac{4}{7}\right)$
17	$-\dfrac{15283}{168} < \dfrac{29}{7}\left(-\dfrac{25}{7}n - \dfrac{19}{3}\right)$
18	$-\dfrac{187}{2} \ge 7\left(-\dfrac{5}{3}n - \dfrac{3}{2}\right) - \dfrac{4}{3}$
19	$\dfrac{3575}{36} > \dfrac{25}{6}\left(\dfrac{8}{3}x + \dfrac{5}{2}\right)$
20	$8k - \dfrac{15}{8}\left(-\dfrac{10}{3}k + 1\right) > -\dfrac{927}{8}$

Graph the solution

Problems

1	$2x - y > -3$ $x - 3y \leq 6$
2	$x + y \leq -3$ $3x - 2y > -4$
3	$3x + 2y \geq 4$ $x - y < 3$
4	$x + y \geq -2$ $x \leq -3$

Writing given a graph

How to solve

Giving a Graph find the solution

Example 1

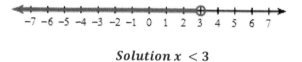

Solution x < 3

Example 2

5)

solution $x \le -5$

Problems

1	–7 –6 –5 –4 –3 –2 –1 0 1 2 3 4 5 6 7
2	–7 –6 –5 –4 –3 –2 –1 0 1 2 3 4 5 6 7
3	–7 –6 –5 –4 –3 –2 –1 0 1 2 3 4 5 6 7
4	–7 –6 –5 –4 –3 –2 –1 0 1 2 3 4 5 6 7
5	–7 –6 –5 –4 –3 –2 –1 0 1 2 3 4 5 6 7
6	–7 –6 –5 –4 –3 –2 –1 0 1 2 3 4 5 6 7
7	–7 –6 –5 –4 –3 –2 –1 0 1 2 3 4 5 6 7
8	–7 –6 –5 –4 –3 –2 –1 0 1 2 3 4 5 6 7
9	–7 –6 –5 –4 –3 –2 –1 0 1 2 3 4 5 6 7
10	–7 –6 –5 –4 –3 –2 –1 0 1 2 3 4 5 6 7

11	—7 —6 —5 —4 —3 —2 —1 0 1 2 3 4 5 6 7
12	—7 —6 —5 —4 —3 —2 —1 0 1 2 3 4 5 6 7
13	—7 —6 —5 —4 —3 —2 —1 0 1 2 3 4 5 6 7
14	—7 —6 —5 —4 —3 —2 —1 0 1 2 3 4 5 6 7
15	—7 —6 —5 —4 —3 —2 —1 0 1 2 3 4 5 6 7
16	—7 —6 —5 —4 —3 —2 —1 0 1 2 3 4 5 6 7
17	—7 —6 —5 —4 —3 —2 —1 0 1 2 3 4 5 6 7
18	—7 —6 —5 —4 —3 —2 —1 0 1 2 3 4 5 6 7
19	—7 —6 —5 —4 —3 —2 —1 0 1 2 3 4 5 6 7
20	—7 —6 —5 —4 —3 —2 —1 0 1 2 3 4 5 6 7

LINEAR FUNCTION

A linear function is any function that can be expressed by one straight line. Our standard form that these functions follow is this:

$$y = f(x) = a + bx$$

Within any given linear function there exists a dependent and an independent variable, our independent variable being x and our dependent variable being y. The a, in this function, is the constant term or the intercept at y (the point at which a line crosses its axis of y). Whenever x = 0, a is our value of the dependent variable. B is the slope or the coefficient of the independent variable.

How to Solve

In order to graph linear functions, we must first find two points that fit the equation, then plot them both, and finally, connect the points with a straight line.

Example 1:

$$y = x + 2$$

First, we find some ordered pairs that would satisfy this function: (-2,0) (-1,1) (0,2) (1,3) (2,4). Now we would only have to plot out our ordered pairs on a graph and the result would be a straight line.

Problems

1	$x\ intercept = 4$ $y\ intercept = 2$
2	$x\ intercept = -4$ $y\ intercept = 1$
3	$y = \dfrac{1}{2}x - 4$
4	$y = -\dfrac{7}{4}x - 3$

SLOPE OF A LINEAR FUNCTION

Our slope within any function that is linear represents the angle at which the line is directed up or downwards. A positive slope indicates that the line is moving up, whereas any negative slope indicates that it is moving down. The vertical change in the line between two points is what is known as their rise, while the horizontal change is what is known as the run.

How to Solve

If two different coordinate pairs are expressed as (x1, y1) as well as (x2, y2), then the equation of the rise would be y2 - y1, while the equation for the run would be x2 - x1.

The slope of a linear function is usually expressed as m and the equation used for finding it is

$$m = \frac{y2 - y1}{x2 - x1}$$

Let's take these two coordinate pairs for example: (3, 6) and (8, 10). To find the rise, we would use this equation: 10 - 6 = 4. To find the run, we would use this equation: 8 - 3 = 5. The slope is expressed as rise / run, so our slope, in this case, would be 4 / 5.

Problems

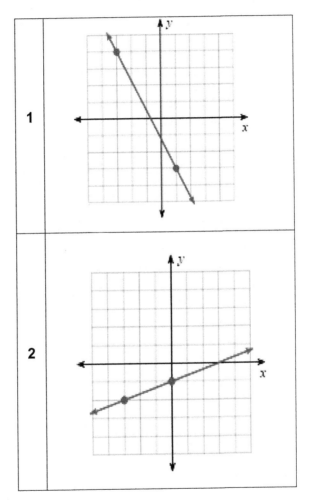

LINEAR INEQUALITIES

Graphing Linear

Next, we should go over how to graph linear inequalities. In order to do this, we must follow three simple steps. The first step is to rearrange the basic equation so that y is placed on the left and everything else is placed on the right. The next step is to plot the y-line. Keep in mind here to use a solid line when the equation states y≤ or y≥, or a dotted line when it states y< or y>. Finally, it is necessary to shade the area above the line in the case of greater than signs (y> and y≥) or shade below the line in the case of less than signs (y< and y≤).

How to solve

Let's learn how to graph these by using the equation y ≤ 2x - 1. Here, the y is already alone on our left side of this inequality, so there is no need to rearrange the order. We would use a solid line in this case because ≤ includes equal to:

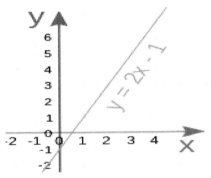

Finally, we need to shade in the area below the line because y is equal to or less than 2x minus 1. This would leave us with this graph:

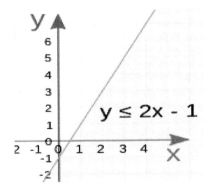

Sketch the graph of each linear inequality

1) $x - 3y \geq 0$

2) $5x + y \geq 2$

Solutions

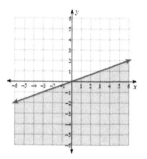 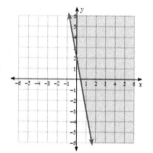

Problems

1	$y \leq -\dfrac{3}{2}x + 2$ $y > \dfrac{1}{2}x - 2$
2	$y > \dfrac{2}{3}x + 3$ $y < -x - 2$
3	$y \geq \dfrac{1}{2}x - 2$ $y \leq 3x + 3$
4	$y < 2x - 1$ $y \leq \dfrac{1}{2}x + 2$

Graphing Systems of Inequalities

How to solve

The easiest and most common way of solving for linear programming problems is to graph our inequalities to create an area called the feasibility region in the y, x plane. To check the solution, it is necessary to determine our coordinates of the corners of the feasibility region and plug them back into our first inequality to determine whether they hold true in it.

Example:

$$\begin{cases} x + 2y \leq 14 \\ 3x - y \geq 0 \\ x - y \leq 2 \end{cases}$$

In here, have to find the minimal and maximal values of

$z = 3x + 4y$ using the inequalities, or constraints, listed above.

These constraints listed above define the feasibility region of our problem. Now, we have to find our points of our corners of our region of feasibility so that we can plug them into the X & Y variables listed within our original equation.

Before we graph these constraints, however, we should first put their equations into more easily graphable versions of themselves:

$$\begin{cases} x+2y \leq 14 \\ 3x-y \geq 0 \\ x-y \leq 2 \end{cases} \Rightarrow \begin{cases} y \leq -\frac{1}{2}x+7 \\ y \leq 3x \\ y \geq x-2 \end{cases}$$

And now it becomes easier to graph our system:

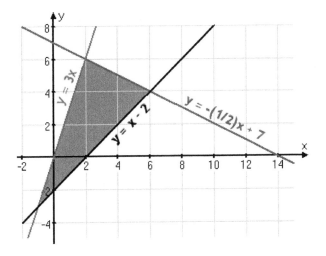

Now, in order to determine our corner points for this region of feasibility, we would need to determine the quantities of all the variables within all the constraints. This procedure here would look like this:

273

$$y = -\frac{1}{2}x + 7$$

$$\frac{1}{2}x + 7 = 3x$$

$$14 = 7x$$

$$2 = x$$

$$y = 3 \cdot 2 = 6$$

Point of the corner at (2,6)

$$y = -\frac{1}{2}x + 7$$

$$y = x - 2$$

$$\frac{1}{2}x + 7 = x - 2$$

$$-x + 14 = 2x - 4$$

$$18 = 3x$$

$$6 = x$$

$$y = 6 - 2 = 4$$

Point of the corner at (6,4)

$$y = 3x$$

$$y = x - 2$$

$$3x = x - 2$$

$$2x = -2$$

$$x = -1$$

$$y = 3(-1) = -3$$

Point of the corner (-1,-3)

As you can see, our corner points of this feasibility region are (2, 6), (6, 4), and (-1, -3).

Our next step is now to determine our maximum and minimum for the system. We would perform this step by plugging in the x and y values of all our corner points to our original equation of

$$z = 3x + 4y$$

- (2, and 6):
$$z = 3(2) + 4(6) = 6 + 24 = 30$$

- (6, and 4):
$$z = 3(6) + 4(4) = 18 + 16 = 34$$

- (−1, and −3):
$$z = 3(-1) + 4(-3) = -3 - 12 = -15$$

Then our maximum for z = 34 happens at (6, and 4)

Problems

1	$y < -x - 2$ $y \leq -6x + 3$
2	$y \geq x - 2$ $y < -\dfrac{1}{3}x + 2$
3	$y \geq 3x - 3$ $y < \dfrac{1}{2}x + 2$
4	$y \geq -4x + 3$ $y \geq 2x - 3$

SOLUTIONS

REFRESH SOME MATH BASIS

Divisibility Rules

Number	divisible by 2, Yes or No
53,764	Yes
1,246	Yes
69,749	No
738	Yes
9,350	Yes
345	No
4,348	Yes
15	No
92,576	Yes
3,273	No

Number	divisible by 3, Yes or No
1,353	Yes
36,696	Yes
4,567	No
35	No
8,241	Yes
56	No
9,132	Yes
25,788	Yes
7,901	No
99	Yes

Number	divisible by 5, Yes or No
75	Yes
450	Yes
4,658	No
81,270	Yes
3,825	Yes
6,200	Yes
12,835	Yes
97	No
1,011	No
52,170	Yes

Number	divisible by 10, Yes or No
100	Yes
3,568	No
40,375	No
785,420	Yes
58,743	No
736,271	No
52,340	Yes
4,890	Yes
34,719	No
89,770	Yes

Least Common Multiple (LCM)

		Find the LCM for the following couples of numbers
1	3, 7	**3**: 3,6,9,12,15,18, **21** **7**: 7,14, **21** $LCM = 21$
2	5, 6	**5**: 5,10,15,20,25,**30** **6**: 6,12,18,24,**30** **LCM = 30**
3	12, 20	**12**: 12,24,36,48, **60** **20**: 20,40,**60** **LCM = 60**
4	4, 6	**4**: 4, 8, **12** **6**: 6, **12** **LCM =12**
5	4, 7	**4**: 4,8,12,16,18,20,24,**28** **7**: 7,14,21,**28** **LCM = 28**

6	12,5	**12**: 12,24,36,48,**60** **5**:5,10,15,20,25,30,35,40,45,50,55,**6,180** **LCM = 60**
7	11,2	**11**: 11,**22** **2**: 2,4,6,8,10,12,14,16,18,20,**22** **LCM = 22**
8	8,12	**8**: 8,16,**24** **12**: 12,**24** **LCM = 24**

Simplify Fractions

1	$\dfrac{15}{5} = \dfrac{15 \div 5}{5 \div 5} = 3$
2	$\dfrac{8}{12} = \dfrac{8 \div 2}{12 \div 2} = \dfrac{4 \div 2}{6 \div 2} = \dfrac{2}{3}$
3	$\dfrac{1}{4} \times 6 = \dfrac{6}{4} = \dfrac{6 \div 2}{4 \div 2} = \dfrac{3}{2}$
4	$\dfrac{\cancel{3}}{\cancel{6}} \times \dfrac{\cancel{6}}{\cancel{9}} = \dfrac{3 \div 3}{6 \div 6} \times \dfrac{6 \div 6}{9 \div 3} = \dfrac{1}{1} \times \dfrac{1}{3} = \dfrac{1}{3}$
5	$\dfrac{\cancel{4}}{12} \times \dfrac{1}{\cancel{8}} = \dfrac{4 \div 4}{12} \times \dfrac{1}{8 \div 4} = \dfrac{1}{12} \times \dfrac{1}{2} = \dfrac{1}{24}$
6	$\dfrac{18}{12} = \dfrac{18 \div 3}{12 \div 3} = \dfrac{6 \div 2}{4 \div 2} = \dfrac{3}{2}$
7	$\dfrac{25}{5} = \dfrac{25 \div 5}{5 \div 5} = 5$
8	$\dfrac{\cancel{4}}{\cancel{3}} \times \dfrac{\cancel{9}}{\cancel{2}} = \dfrac{4 \div 2}{3 \div 3} \times \dfrac{9 \div 3}{2 \div 2} = 2 \times 3 = 6$

Operations with Fractions

1	$\dfrac{2}{7} + \dfrac{5}{2} = \dfrac{2(14 \div 7) + 5(14 \div 2)}{14} = \dfrac{4 + 35}{14} = \dfrac{39}{14}$
2	$\dfrac{6}{5} \div \dfrac{3}{20} = \dfrac{6}{5} \cdot \dfrac{20}{3} = \dfrac{6 \div 3}{5 \div 5} \cdot \dfrac{20 \div 5}{3 \div 3} = 2 \cdot 4 = 8$
3	$\dfrac{30}{4} = \dfrac{30 \div 2}{4 \div 2} = \dfrac{15}{2}$
4	$\dfrac{7}{8} - \dfrac{3}{4} = \dfrac{7(32 \div 8) - 3(32 \div 4)}{32} = \dfrac{28 - 24}{32} = \dfrac{4}{32} = \dfrac{1}{8}$
5	$\dfrac{7}{5} \times \dfrac{20}{21} = \dfrac{7 \div 7}{5} \times \dfrac{20 \div 5}{21 \div 7} = \dfrac{1}{5} \cdot \dfrac{4}{3} = \dfrac{4}{15}$
6	$\dfrac{2}{5} \div \dfrac{3}{15} = \dfrac{2}{5} \times \dfrac{15}{3} = \dfrac{2}{5 \div 5} \times \dfrac{15 \div 5}{3} = 2 \times 1 = 2$

7	$\dfrac{9}{3} - \dfrac{3}{12} = \dfrac{9(12 \div 3) - 3(12 \div 12)}{12} = \dfrac{36 - 3}{12} = \dfrac{33}{12} = \dfrac{11}{4}$
8	$\dfrac{8}{5} + \dfrac{3}{4} = \dfrac{8(20 \div 5) + 3(20 \div 4)}{20} = \dfrac{32 + 15}{20} = \dfrac{47}{20}$
9	$\dfrac{40}{12} = \dfrac{40 \div 2}{12 \div 2} = \dfrac{20}{6} = \dfrac{20 \div 2}{6 \div 2} = \dfrac{10}{3}$
10	$\dfrac{2}{3} \div \dfrac{3}{15} = \dfrac{2}{3} \times \dfrac{15}{3} = \dfrac{2}{3 \div 3} \times \dfrac{15 \div 3}{3} = 2 \times \dfrac{5}{3} = \dfrac{10}{3}$
11	$\dfrac{90}{10} = \dfrac{90 \div 10}{10 \div 10} = 9$

INTEGERS

Adding and Subtracting

1	-5	**11**	$\dfrac{11}{20}$
2	-5	**12**	$-\dfrac{121}{21}$
3	9	**13**	$-\dfrac{77}{24}$
4	0	**14**	$-\dfrac{10}{3}$
5	9	**15**	$\dfrac{21}{8}$
6	-1	**16**	$-\dfrac{19}{5}$
7	-9	**17**	$\dfrac{7}{6}$
8	1	**18**	$-\dfrac{141}{28}$
9	-2	**19**	$\dfrac{99}{56}$
10	-5	**20**	$-\dfrac{143}{28}$

Multiplying

1	-24	**11**	$-\dfrac{28}{5}$
2	144	**12**	-1
3	350	**13**	$\dfrac{140}{27}$
4	-54	**14**	$-\dfrac{13}{30}$
5	8	**15**	$\dfrac{5}{2}$
6	490	**16**	$-\dfrac{133}{45}$
7	-126	**17**	$-\dfrac{19}{40}$
8	240	**18**	$-\dfrac{22}{15}$
9	-48	**19**	$-\dfrac{35}{12}$
10	-140	**20**	$-\dfrac{68}{25}$

Dividing

1	-2	**10**	$\dfrac{135}{4}$
2	9	**11**	$\dfrac{14}{3}$
3	-5	**12**	$-\dfrac{26}{5}$
4	-3	**13**	$\dfrac{5}{9}$
5	2	**14**	$-\dfrac{85}{28}$
6	-8	**15**	$\dfrac{15}{14}$
7	8	**16**	$-\dfrac{28}{31}$
8	-7	**17**	$\dfrac{12}{11}$
9	10	**18**	$-\dfrac{5}{8}$

POWERS AND EXPONENTS

Product property

1	8	**11**	$2x^6$
2	$8v$	**12**	$6x^3$
3	$8a$	**13**	$3v^2$
4	$16n^7$	**14**	$\dfrac{48}{n^2}$
5	$\dfrac{12}{n}$	**15**	$\dfrac{48}{n^7}$
6	$6v^5$	**16**	$3x^8$
7	1	**17**	$\dfrac{36}{p}$
8	$3n$	**18**	$\dfrac{3}{x^2}$
9	$3m^2$	**19**	$12k^3$
10	$12x^2$	**20**	$16k^4$

Power Property

1	$27x^6$	**11**	$27x^6$
2	v^9	**12**	$81v^{12}$
3	1	**13**	x^5
4	x^6	**14**	$-\dfrac{1}{x^{12}}$
5	$8b^6$	**15**	$16x^2$
6	$9m^6$	**16**	$\dfrac{p^{15}}{8}$
7	$27x^3$	**17**	1
8	$9k^2$	**18**	$\dfrac{125}{n^{15}}$
9	$27b^9$	**19**	$16x^2$
10	$4m^4$	**20**	$25m^6$

Quotient Property

1	m^3	**11**	$-\dfrac{1}{2}$
2	$\dfrac{1}{3r}$	**12**	$-\dfrac{n^3}{2}$
3	$\dfrac{2}{3k}$	**13**	$\dfrac{5}{k^8}$
4	$\dfrac{x^2}{3}$	**14**	$\dfrac{5}{2}$
5	$\dfrac{2}{x}$	**15**	$\dfrac{3}{4n^3}$
6	$\dfrac{1}{n^2}$	**16**	$\dfrac{2}{n^8}$
7	$\dfrac{1}{x}$	**17**	$\dfrac{4x}{5}$
8	$\dfrac{2}{3p}$	**18**	$-\dfrac{4m^3}{5}$
9	$2a^2$	**19**	$\dfrac{5r^3}{2}$
10	$\dfrac{2}{3}$	**20**	$\dfrac{1}{p^3}$

ALGEBRIC OPERATIONS IN THE CORRECT ORDER

1	1	11	51
2	36	12	-2
3	3	13	-11
4	2	14	14
5	3	15	6
6	$\dfrac{15}{4}$	16	$\dfrac{175}{1152}$
7	$\dfrac{37}{24}$	17	$\dfrac{1057}{40}$
8	$\dfrac{12}{5}$	18	$-\dfrac{341}{90}$
9	$\dfrac{29}{12}$	19	$-\dfrac{17}{12}$
10	$\dfrac{27}{4}$	20	$\dfrac{29}{36}$

EVALUATION EXPRESSIONS

1	7	11	$\dfrac{3481}{144}$
2	3	12	$\dfrac{49}{4}$
3	8	13	$\dfrac{215}{24}$
4	10	14	$\dfrac{15}{4}$
5	2	15	$\dfrac{21}{16}$
6	6	16	1
7	7	17	$\dfrac{13}{6}$
8	36	18	$\dfrac{3}{2}$
9	6	19	2
10	1	20	$\dfrac{242}{21}$

POLYNOMIALS
Multiplying Polynomials

1	$36a - 18$	**11**	$30p^2 - 13p - 56$
2	$24r + 28$	**12**	$16b^2 + 32b + 7$
3	$28x^2 + 56x$	**13**	$15n^2 + 38n + 7$
4	$42p + 49$	**14**	$m^2 - 12m + 32$
5	$24n^2 + 56n$	**15**	$32r^2 - 52r + 21$
6	$\dfrac{99}{20}p^2 - \dfrac{231}{20}p$	**16**	$\dfrac{247}{18}x^2 - \dfrac{148}{21}x + \dfrac{6}{7}$
7	$v + \dfrac{1}{4}$	**17**	$4x^2 + \dfrac{1}{4}x - \dfrac{351}{32}$
8	$\left(\dfrac{7}{12}v + \dfrac{1}{2}\right)$	**18**	$\dfrac{153}{40}n^2 - \dfrac{543}{80}n + 3$
9	$\left(\dfrac{17}{9}n + \dfrac{17}{14}\right)$	**19**	$\dfrac{34}{5}k^2 - \dfrac{331}{60}k - \dfrac{12}{5}$
10	$\left(\dfrac{45}{28}p + \dfrac{115}{56}\right)$	**20**	$\dfrac{1}{3}r^2 + \dfrac{59}{24}r + \dfrac{11}{4}$

Adding and Subtracting

1	$5n^3 + 4n^2 + 4n$	**11**	$-\dfrac{123}{35}n^3 + \dfrac{49}{15}n$
2	$5n^4 + 8n^2$	**12**	$\dfrac{12}{5}k^4 + \dfrac{11}{4}$
3	$-v^3 - 3v^2$	**13**	$\dfrac{38}{35}b^4 + \dfrac{23}{14}b^2$
4	$-n^2 + 6n$	**14**	$-\dfrac{17}{5}r^4 + \dfrac{77}{6}r^3$
5	$8n^4 + 7n^2$	**15**	$\dfrac{11}{6}b^3 - \dfrac{127}{20}b$
6	3	**16**	$\dfrac{9}{5}m^4 + \dfrac{93}{35}m^3 + \dfrac{15}{4}$
7	$8v^4 - 3$	**17**	$\dfrac{9}{7}a^2 + \dfrac{56}{15}a$
8	$16b^2$	**18**	$\dfrac{71}{40}k^4 - \dfrac{17}{20}k^3$
9	$2x^3 - 3x^2$	**19**	$-\dfrac{1}{6}n^3 - \dfrac{1}{2}n$
10	$6v^2 - 1$	**20**	$-\dfrac{104}{21}n^4 + \dfrac{101}{35}n^3$

Dividing

1	$v + 9$	11	$5n + 3$
2	$r - 9$	12	$r - 10$
3	$a - 1$	13	$n - 5$
4	$b - 2$	14	$m - 9$
5	$k - 3$	15	$v + 2$
6	$m + 2$	16	$x + 1$
7	$n - 1$	17	$n - 1$
8	$4k + 6$	18	$x + 10$
9	$4n - 2$	19	$x - 2$
10	$9x - 3$		

BINOMIALS

Square of Binomials

1	$b^2 - 4b + 4$	**11**	$9x^2 - 42x + 49$
2	$n^2 - 8n + 16$	**12**	$4x^2 + 8x + 4$
3	$n^2 - 1$	**13**	$36n^2 - 96n + 64$
4	$r^2 + 10r + 25$	**14**	$16b^2 + 16b + 4$
5	$x^2 - 4$	**15**	$m^2 + 6m + 9$
6	$x^2 - \dfrac{10}{3}x + \dfrac{25}{9}$	**16**	$4x^2 + \dfrac{2}{3}x + \dfrac{1}{36}$
7	$x^2 - x + \dfrac{1}{4}$	**17**	$9 - 10x + \dfrac{25}{9}x^2$
8	$k^2 - \dfrac{12}{5}x + \dfrac{36}{25}$	**18**	$\dfrac{1}{16}x^4 + \dfrac{29}{16}x^2 + \dfrac{841}{64}$
9	$a^2 - \dfrac{14}{5}a + \dfrac{49}{25}$	**19**	$1 + \dfrac{3}{2}p + \dfrac{9}{16}p^2$
10	$k^2 - \dfrac{18}{5}k + \dfrac{81}{25}$	**20**	$\dfrac{9}{16}x^6 + \dfrac{3}{4}x^3 + \dfrac{1}{4}$

Differences of two Squares

1	$1 - 64p^2$	**11**	$\dfrac{1}{9}x^2 - \dfrac{625}{64}$
2	$36x^2 - 36$	**12**	$-1 + \dfrac{64}{25}n^2$
3	$49k^2 - 9$	**13**	$\dfrac{25}{49} - \dfrac{169}{16}p^2$
4	$16x^2 - 1$	**14**	$\dfrac{225}{64} - \dfrac{961}{64}n^2$
5	$49x^2 - 25$	**15**	$\dfrac{1}{16}m^2 - \dfrac{9}{4}$
6	$49x^2 - 4$	**16**	$-4v^2 + \dfrac{4}{9}$
7	$9 - 9b^2$	**17**	$\dfrac{9}{4}n^2 - \dfrac{49}{25}$
8	$49x^2 - 25$	**18**	$-4 + \dfrac{484}{25}x^2$
9	$64x^2 - 4$	**19**	$n^2 - \dfrac{121}{25}$
10	$4p^2 - 64$	**20**	$-1 + \dfrac{81}{64}x^2$

FACTORING

Factoring By grouping

1	$(b^2 + 3)(2b - 5)$
2	$(x^2 + 4)(x - 5)$
3	$(3x^2 + 1)(2x - 5)$
4	$(2p^2 + 5)(4p - 5)$
5	$(2x^2 + 1)(2x - 5)$
6	$(3x^2 + 4)(x - 4)$
7	$(3r^2 + 2)(3r + 4)$
8	$(r^2 + 5)(r - 4)$
9	$(5k^2 + 2)(4k + 5)$
10	$(3r^2 + 2)(5r + 2)$

11	$2(4n^2 + 3)(3n - 2)$
12	$2(5n^2 + 3)(n - 4)$
13	$2(5x^2 + 3)(x - 2)$
14	$2(3r^2 + 4)(3r + 1)$
15	$4(2k^2 + 3)(k - 1)$
16	$(3n^2 + 2)(3n + 8)$
17	$(8x^2 + 5)(4x + 1)$
18	$(2n^2 + 1)(4n - 1)$
19	$3(8x^2 + 5)(x + 8)$
20	$2(7x^2 - 5)(3x - 7)$

Common Factor only

1	$2(9b^{10} + 7b^2 - 5)$
2	$8b^2(10b^3 - 5b^2 + 8)$
3	$8(-7 - 7v + 2v^3)$
4	$4v^3(5v^8 + 9v + 2)$
5	$7(3x^2 + 8x + 9)$
6	$6r(7r^5 + 7r^2 - 8)$
7	$5x^4(3x^3 - 4x + 3)$
8	$10(-7x^5 + 5x - 8)$
9	$6m^3(-3m^5 + m^2 - 5)$
10	$2p(5p^3 - 7p - 3)$

11	$$9x^2(5x^3 + 10x + 2)$$
12	$$10a^2(-5a^3 + 9a + 9)$$
13	$$4(6n - 9 + 7n^4)$$
14	$$9(-6 + 4r - r^3)$$
15	$$6(2n^3 + 4n + 5)$$
16	$$x(-4 + 3x + 5x^3)$$
17	$$2n^2(6n^3 + 9n - 7)$$
18	$$-7b(4b^2 + 9b + 6)$$
19	$$7(-9 + 10p^2 - 8p^4)$$
20	$$-6n^3(n^3 + 7n + 1)$$

Greatest Common Factor GCF

1) $3x^3 + 27x^2 + 9x$

$$Factors\ of\ 3x^3 = \mathbf{3} \cdot \mathbf{x} \cdot x \cdot x$$
$$Factors\ of\ 27x^2 = \mathbf{3} \cdot 3 \cdot 3 \cdot \mathbf{x} \cdot x$$
$$Factors\ of\ 9x = 3 \cdot \mathbf{3} \cdot \mathbf{x}$$
$$GCF = \mathbf{3x}$$

2) $36x^2 - 64y^4$

$$Factors\ of\ 36x^2 = \mathbf{2} \cdot \mathbf{2} \cdot 3 \cdot 3 \cdot x \cdot x$$
$$Factors\ of\ 64y^4 = \mathbf{2} \cdot \mathbf{2} \cdot 2 \cdot 2 \cdot 2 \cdot 2 \cdot y \cdot y \cdot y \cdot y$$
$$GCF = \mathbf{4}$$

3) Find GFC between 81 and 180

$$Factors\ of\ 81 = \mathbf{3} \cdot \mathbf{3} \cdot 3 \cdot 3$$
$$Factors\ of\ 180 = \mathbf{3} \cdot \mathbf{3} \cdot 2 \cdot 2 \cdot 5$$
$$GCF = 3 \cdot 3 = \mathbf{9}$$

4) Find GFC of the following polynomial:

$$6x^2y + 14xy^2 - 42xy - 2x^2y^2$$

$$Factors\ of\ 6x^2y:\ \mathbf{2} \cdot 3 \cdot \mathbf{x} \cdot x \cdot \mathbf{y}$$
$$Factors\ of\ 14xy^2:\ \mathbf{2} \cdot 7 \cdot \mathbf{x} \cdot \mathbf{y} \cdot y$$
$$Factors\ of\ 42xy:\ \mathbf{2} \cdot 3 \cdot 7 \cdot \mathbf{x} \cdot \mathbf{y}$$
$$Factors\ of\ 2x^2y^2:\ \mathbf{2} \cdot \mathbf{x} \cdot x \cdot \mathbf{y} \cdot y$$

$$GCF = \mathbf{2xy}$$

QUADRATIC EXPRESSION

Factoring Quadratic Polynomials

1	$(a - 10)^2$	11	$b(7b - 8)$
2	$(b - 9)(b + 10)$	12	$3(5a + 3)(a - 5)$
3	$(p + 2)(p - 4)$	13	$(7n + 2)(n + 8)$
4	$(n - 1)(n - 3)$	14	$(2x + 3)(x + 7)$
5	$(x - 8)(x + 4)$	15	$2(7n - 10)(n - 7)$
6	$(n + 9)(n + 5)$	16	$4(5k - 2)(k + 10)$
7	$(x - 9)(x - 6)$	17	$5(5k + 9)(k - 6)$
8	$3(b + 5)(b - 10)$	18	$2(3a + 8)(a + 2)$
9	$5(m + 8)(m + 3)$	19	$4(2x - 5)(x + 9)$
10	$(n - 9)(n + 2)$	20	$4(5a - 3)(a - 4)$

RADICAL EXPRESSIONS

Adding and Subtracting

1	$-\sqrt{2}$	**11**	$-\sqrt{2} - 5\sqrt{5}$
2	$-6\sqrt{3}$	**12**	$\sqrt{5} - \sqrt{6}$
3	$15\sqrt{2}$	**13**	$-\sqrt{3}$
4	$2\sqrt{2}$	**14**	$5\sqrt{5} - \sqrt{6}$
5	0	**15**	$3\sqrt{5} + \sqrt{2}$
6	$-11\sqrt{5}$	**16**	$\sqrt{3} + 2\sqrt{5}$
7	$6\sqrt{6}$	**17**	$4\sqrt{3}$
8	$-4\sqrt{6}$	**18**	$5\sqrt{3} - 2\sqrt{5}$
9	$-8\sqrt{5}$	**19**	$5\sqrt{6}$
10	$-12\sqrt{3}$	**20**	$5\sqrt{3} - 2\sqrt{6}$

Dividing

1	$\dfrac{-20\sqrt{3}-15}{13}$	**11**	$\dfrac{-3-\sqrt{2}}{7}$
2	$\dfrac{1+\sqrt{5}}{2}$	**12**	$\dfrac{16+4\sqrt{3}}{13}$
3	$\dfrac{-3-6\sqrt{3}}{22}$	**13**	$-3\sqrt{2}+4$
4	$\dfrac{-5\sqrt{5}+2\sqrt{3}}{113}$	**14**	$4\sqrt{2}+2\sqrt{6}$
5	$\dfrac{5-\sqrt{3}}{11}$	**15**	$\dfrac{4+\sqrt{2}}{14}$
6	$\dfrac{-15\sqrt{2}-5\sqrt{10}}{4}$	**16**	$\dfrac{-2\sqrt{5}+4\sqrt{3}}{7}$
7	$\dfrac{-10-15\sqrt{5}}{41}$	**17**	$\dfrac{-4+6\sqrt{2}}{7}$
8	$\dfrac{-5\sqrt{3}+4\sqrt{15}}{55}$	**18**	$\dfrac{-1-\sqrt{5}}{4}$
9	$\dfrac{-3\sqrt{5}+2\sqrt{15}}{3}$	**19**	$\dfrac{2+4\sqrt{3}}{11}$
10	$\dfrac{3+\sqrt{5}}{2}$	**20**	$\dfrac{20-5\sqrt{3}}{13}$

Multiplying

1	$2\sqrt{2}$	**11**	$5 + 3\sqrt{5}$
2	$5\sqrt{15}$	**12**	$-5\sqrt{2} - 5\sqrt{5}$
3	$10\sqrt{3}$	**13**	$3 + 3\sqrt{3}$
4	$\sqrt{15}$	**14**	$-8\sqrt{5} - 20\sqrt{10}$
5	$3\sqrt{2}$	**15**	$15\sqrt{5} - 25$
6	$45\sqrt{6}$	**16**	$5 + 5\sqrt{5}$
7	$-20\sqrt{15}$	**17**	$3\sqrt{3} + 3$
8	100	**18**	$-15\sqrt{5} + 20\sqrt{30}$
9	-500	**19**	$4\sqrt{3} + \sqrt{30}$
10	$40\sqrt{3}$	**20**	$-5\sqrt{3} - 3\sqrt{2}$

Simplify Radical Expressions

1	$-72\sqrt{x}$	11	$2\sqrt{5x}$
2	$-48\sqrt{7x}$	12	$2b\sqrt{5}$
3	$-8n\sqrt{2}$	13	$4n\sqrt{2}$
4	$-25k\sqrt{3k}$	14	$\sqrt{30n}$
5	$20n\sqrt{5n}$	15	$4m$
6	$10\sqrt{30}$	16	$-100p^2\sqrt{2}$
7	$-6\sqrt{210n}$	17	$6\sqrt{42}$
8	$45x\sqrt{5}$	18	$-10\sqrt{105}$
9	$80\sqrt{7x}$	19	$-6x\sqrt{2}$
10	$-30b^2\sqrt{5}$	20	$4\sqrt{210a}$

RATIONAL EXPRESSIONS

Adding and Subtracting

1	$\dfrac{n^2 - n + 2}{2(n-3)}$	**11**	$\dfrac{9x + 32}{2x(x+5)}$
2	$\dfrac{2x + 16}{3(x+6)}$	**12**	$\dfrac{k^2 - 5k - 18}{3k(k+3)}$
3	$\dfrac{-6r + 7}{3r - 1}$	**13**	$\dfrac{-4a^3 + 16a^2 - 11a + 6}{(a-3)(a-1)}$
4	$\dfrac{6m - 3 + m^2}{m(m-3)}$	**14**	$\dfrac{36x + x^2 - 16}{6(x-4)}$
5	$\dfrac{23x + 14}{2(5x+4)}$	**15**	$\dfrac{63x + 6 + 6x^3 + 46x^2}{6x(3x+5)}$
6	$\dfrac{12x + 8 + 10x^3}{5x^2(3x+2)}$	**16**	$\dfrac{-2b + 20}{3(b+5)}$
7	$\dfrac{5n^2 - 25n - 6}{3(n-5)}$		
8	$\dfrac{21r - 12 - 3r^2}{(r-1)(3r-2)}$		
9	$\dfrac{10x^2 + 5x}{(x-1)(x+2)}$		
10	$\dfrac{9r}{(r-5)(r-2)}$		

Dividing

Please see previous chapter "The Remainder Theorem Method in the Polynomials chapter.

Multiplying

1	$\dfrac{(v-6)(v+2)}{9}$	**11**	$\dfrac{n-9}{8n^2}$
2	$\dfrac{2}{(a+1)(a-9)}$	**12**	$\dfrac{x+6}{2x^2}$
3	$\dfrac{3}{2}$	**13**	$-r+3$
4	$\dfrac{x-8}{x+9}$	**14**	$\dfrac{-p-5}{p+3}$
5	$\dfrac{x+8}{9x}$	**15**	$\dfrac{x-8}{x+9}$
6	$\dfrac{10}{(k-8)(k+3)}$	**16**	$\dfrac{n-10}{n+4}$
7	$\dfrac{p+4}{8}$	**17**	$\dfrac{x-9}{9x(x-3)}$
8	$\dfrac{x-5}{x+5}$	**18**	$\dfrac{p-6}{2}$
9	$\dfrac{a-6}{a-9}$	**19**	$\dfrac{x+1}{7}$
10	$x-3$	**20**	$-(x-3)$

Simplifying

1	$\dfrac{1}{n+5}$	**11**	$r+1$
2	$\dfrac{1}{n+2}$	**12**	$5n$
3	$r+4$	**13**	$\dfrac{3x+2}{4}$
4	$r+3$	**14**	$\dfrac{1}{3v}$
5	$\dfrac{3x-2}{4}$	**15**	$\dfrac{1}{a+1}$
6	$\dfrac{3m}{5(m-1)}$	**16**	$\dfrac{4v}{5v+1}$
7	$a+2$	**17**	$\dfrac{1}{3}$
8	$\dfrac{4}{3x+2}$	**18**	$\dfrac{x+1}{2x}$
9	$\dfrac{1}{p-2}$	**19**	$\dfrac{1}{5}$
10	$n-2$	**20**	$\dfrac{3}{2(n+1)}$

ONE STEP EQUATION

1	2	**11**	-22
2	3	**12**	550
3	9	**13**	15
4	-7	**14**	96
5	6	**15**	-14
6	$\dfrac{9}{8}$	**16**	$-\dfrac{2}{15}$
7	$-\dfrac{23}{7}$	**17**	$\dfrac{50}{7}$
8	$\dfrac{13}{6}$	**18**	$\dfrac{7}{15}$
9	$\dfrac{3}{2}$	**19**	$\dfrac{19}{23}$
10	$\dfrac{57}{10}$	**20**	$-\dfrac{67}{20}$

MULTI STEP EQUATION

1	-8	11	8
2	4	12	$-\dfrac{27}{7}$
3	6	13	7
4	-4	14	$\dfrac{31}{8}$
5	-8	15	6
6	-7	16	$\dfrac{14}{3}$
7	7	17	8
8	-8	18	5
9	5	19	8
10	7	20	-8

EQUATIONS WITH ABSOLUTE VALUE

1	14, 2	**11**	$\dfrac{37}{10}, -\dfrac{1}{30}$
2	$-4, -14$	**12**	$\dfrac{9}{5}, -\dfrac{9}{5}$
3	$35, -35$	**13**	$-\dfrac{9}{8}, \dfrac{9}{8}$
4	$6, -6$	**14**	2, 1
5	$10, -10$	**15**	$-\dfrac{18}{5}, \dfrac{18}{5}$
6	$12, -12$	**16**	$\dfrac{23}{8}, \dfrac{49}{72}$
7	13, 1	**17**	$\dfrac{27}{10}, \dfrac{3}{10}$
8	$0, -8$	**18**	$\dfrac{17}{10}, -\dfrac{17}{10}$
9	$1, -11$	**19**	$\dfrac{7}{6}, -\dfrac{7}{6}$
10	$4, -2$	**20**	$0, -\dfrac{7}{3}$

QUADRATIC EQUATIONS

Solve by Factoring

1	$-8, 0$	**11**	$7, 1$
2	$-3, 0$	**12**	$-1, -4$
3	$8, 0$	**13**	-4
4	$5, 3$	**14**	$-5, 5$
5	$-5, 2$	**15**	$-3, -6$
6	$8, -1$	**16**	$-7, 2$
7	$-5, 6$	**17**	$3, 7$
8	$-8, -3$	**18**	$-4, 0$
9	$2, 6$	**19**	$8, 2$
10	-8	**20**	$-2, 0$

Solving by Quadratic Formula

1	$4, -3$	**11**	$\dfrac{5}{2}, -1$
2	*no solution*	**12**	$6, -2$
3	$3, -\dfrac{7}{2}$	**13**	$3, -4$
4	$2, 1$	**14**	$1, -\dfrac{3}{2}$
5	$4, -3$	**15**	$2, -\dfrac{1}{2}$
6	$2, -3$	**16**	$1, -4$
7	$2, -4$	**17**	$\dfrac{3}{2}, -2$
8	$4, -1$	**18**	$6, -4$
9	$1, \dfrac{1}{2}$	**19**	$1, -\dfrac{7}{2}$
10	$\dfrac{3}{2}, -1$	**20**	$5, -4$

Solving by taking square roots

1	$5, -5$	**11**	$\sqrt{77}, -\sqrt{77}$
2	*no solution*	**12**	$\sqrt{6}, -\sqrt{6}$
3	$\sqrt{86}, -\sqrt{86}$	**13**	*no solution*
4	$6, -6$	**14**	$3\sqrt{10}, -3\sqrt{10}$
5	$7, -7$	**15**	$10, -10$
6	*no solution*	**16**	$4, -4$
7	$5\sqrt{2}, -5\sqrt{2}$	**17**	$\sqrt{29}, -\sqrt{29}$
8	$2\sqrt{17}, -2\sqrt{17}$	**18**	$\sqrt{62}, -\sqrt{62}$
9	$\sqrt{35}, -\sqrt{35}$	**19**	$3\sqrt{7}, -3\sqrt{7}$
10	*no solution*	**20**	*no solution*

RADICAL EQUATIONS

1	-1	11	$2, 7$
2	0	12	4
3	4	13	$0, 2$
4	30	14	$8, 4$
5	16	15	1
6	10	16	6
7	*no solution*	17	$0, 1$
8	7	18	$8, 5$
9	5	19	5
10	8	20	6

RATIONAL EQUATIONS

1	7	**11**	$-\dfrac{31}{4}$
2	$\dfrac{8}{3}$	**12**	$1, -2$
3	-1	**13**	$-\dfrac{16}{3}$
4	-2	**14**	-4
5	-8	**15**	$-2, 7$
6	-2	**16**	$\dfrac{4}{3}$
7	4	**17**	2
8	$\dfrac{11}{3}$	**18**	3
9	7	**19**	-4
10	1	**20**	5

LINEAR EQUATIONS

1)

$$3(x + 5) = 2(-6 - x) - 2x$$
$$3x + 15 = -12 - 2x - 2x =$$
$$3x + 2x + 2x + 15 + 12 = 0$$
$$7x + 27 = 0$$
$$7x = -27$$
$$x = -\frac{27}{7}$$

2)

$$-(-3 + x) + 4 = 2(x + 3) - 6x + 7$$
$$+3 - x + 4 = 2x + 6 - 6x + 7$$
$$-x - 2x = -3 - 4 + 6 + 7$$
$$-3x = 6$$
$$3x = -6$$
$$x = -\frac{6}{3} = -2$$

3)

$$-\frac{3}{4} - \frac{1}{2}x + \frac{3}{4} = \frac{5}{6} - \frac{3}{4}$$
$$-\frac{1}{2}x = \frac{3}{4} - \frac{3}{4} + \frac{5}{6} - \frac{3}{4}$$
$$-\frac{1}{2}x = \frac{9 - 9 + 10 - 9}{12} = \frac{1}{12}$$
$$-\frac{1}{2}x = \frac{1}{12}$$

since the sign of x is negative you can change it
by changing the signs in both sides

$$\frac{1}{2}x = -\frac{1}{12}$$

now to find x you divide $-\frac{1}{12}$ *by* $\frac{1}{2}$ *that is to multiply*

$-\frac{1}{2}$ *with the opposite of* $\frac{1}{2}$ *that is 2*

$$x = -\frac{1}{12} \cdot \frac{2}{1} = -\frac{2}{12}$$

remember to simplify the fraction when it is possible
in this case 2 and 12 can be divided each one by 2

$$x = -\frac{2}{12} = -\frac{1}{6}$$

4)

$$\frac{m-2}{3} + 1 = \frac{2m}{7}$$

$$\frac{m-2}{3} - \frac{2m}{7} = -1$$

$$\frac{7(m-2) - 3(2m)}{21} = -1$$

$$\frac{7m - 14 - 6m}{21} = -1$$

$$\frac{m - 14}{21} = -1$$

$$m - 14 = -1 \cdot (21)$$

$$m = 14 - 21$$

$$\boldsymbol{m = -7}$$

Problems with variable in the denominator

5)

$$\frac{2x}{x+3} = \frac{3}{x-10} + 2$$

The first step is to factor the denominators to get the LCM that is (x+3)(x-10)

In this case we have the variable x in the denominator, so we need to avoid x=−3 and x= −10 because we will divide by 0 that is not possible.

In this condition, to solve the equation, we can multiply the LCM in both sides, so

$$(x+3)(x-10) \cdot \frac{2x}{x+3} = \left(\frac{3}{x-10} + 2\right) \cdot (x+3)(x-10)$$

$$(x-10)2x = \left(\frac{3 + 2x - 20}{x - 10}\right) \cdot (x+3)(x-10).$$

$$(x-10)2x = \left(\frac{2x - 17}{x - 10}\right) \cdot (x+3)(x-10).$$

$$2x^2 - 20x = \frac{2x - 17}{x - 10} \cdot (x+3)(x-10)$$

$$2x^2 - 20x = (2x - 17)(x+3) =$$

$$2x^2 - 20x = 2x^2 + 6x - 17x - 51$$

$$2x^2 - 2x^2 - 20x - 6x + 17x - 51$$

$$-26x + 17x = -51$$

$$-9x = -51$$

$$9x = 51$$

$$x = \frac{51}{9} = \frac{17}{3}$$

6)

$$\frac{2}{x + 2} = \frac{-x}{x^2 + 5x + 6}$$

The first step is to factor the denominators to get the LCM that is

$$(x + 2)(x + 3)$$

In this case we have the variable x in the denominator, so we

need to avoid $x = -2$ and x

$= -3$ because we will divide by 0 that is not possible.

In this condition, to solve the equation, we can multiply the

LCM in both sides, so

$$\cancel{(x + 2)}(x + 3) \cdot \frac{2}{\cancel{x + 2}} = \cancel{(x + 2)(x + 3)} \cdot \frac{-x}{\cancel{(x + 2)(x + 3)}}$$

$$(x + 3)2 = -x$$

$$2x + 6 = -x$$

$$2x + x = -6$$

$$3x = -6$$

$$x = -\frac{6}{3} = -2$$

EQUATION SYSTEMS

Substitution Method

1	$(-2, -4)$	**11**	$(-1, 1)$
2	$(1, -1)$	**12**	$(-1, -3)$
3	$(3, -3)$	**13**	$(-3, -2)$
4	$(2, -1)$	**14**	$(-1, 3)$
5	$(1, 4)$	**15**	*no solution*
6	$(1, 4)$	**16**	$(2, -1)$
7	$(1, 4)$	**17**	$(-2, -4)$
8	$(-4, -1)$	**18**	*no solution*
9	$(4, -4)$	**19**	$(1, -4)$
10	$(4, -1)$	**20**	*no solution*

Elimination Method

1	$(-1, -3)$	**11**	$(1, -3)$
2	$(1, -2)$	**12**	*no solution*
3	$(4, 2)$	**13**	*no solution*
4	$(-3, 1)$	**14**	$(-4, 3)$
5	$(-4, -4)$	**15**	*no solution*
6	$(-2, -2)$	**16**	$(2, -3)$
7	$(3, 1)$	**17**	$(-3, 4)$
8	$(-2, -3)$	**18**	$(1, 4)$
9	$(-3, -3)$	**19**	$(-4, 3)$
10	$(-2, 2)$	**20**	$(-4, 2)$

INEQUALITIES
Absolute Value

1	$v > 12 \text{ or } v < 2$
2	$a \geq 13 \text{ or } a \leq -5$
3	$b \geq 25 \text{ or } b \leq -25$
4	$1 < r < 11$
5	$v \geq 12 \text{ or } v \leq -2$
6	$-\dfrac{5}{7} \leq p \leq \dfrac{5}{7}$
7	$\dfrac{7}{4} \leq p \leq \dfrac{35}{12}$
8	$-\dfrac{3}{7} \leq n \leq \dfrac{3}{7}$
9	$x \geq -\dfrac{17}{5} \text{ or } x \leq -\dfrac{127}{20}$
10	$n > \dfrac{7}{2} \text{ or } n < -\dfrac{19}{4}$

Compound

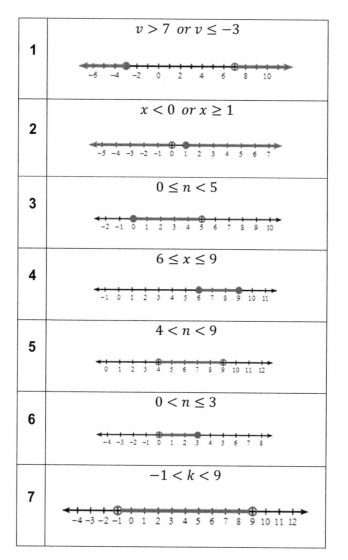

1	$v > 7 \ or \ v \le -3$
2	$x < 0 \ or \ x \ge 1$
3	$0 \le n < 5$
4	$6 \le x \le 9$
5	$4 < n < 9$
6	$0 < n \le 3$
7	$-1 < k < 9$

8	$-4 \le a \le 0$ 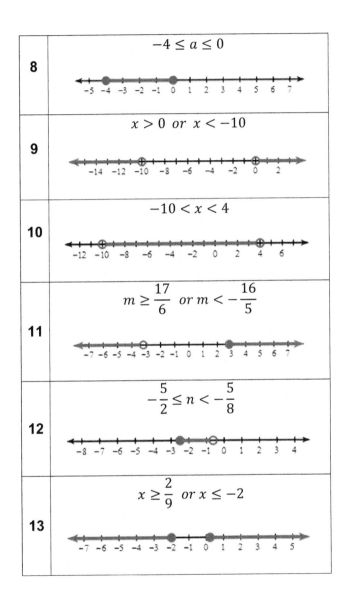
9	$x > 0 \ or \ x < -10$
10	$-10 < x < 4$
11	$m \ge \dfrac{17}{6} \ or \ m < -\dfrac{16}{5}$
12	$-\dfrac{5}{2} \le n < -\dfrac{5}{8}$
13	$x \ge \dfrac{2}{9} \ or \ x \le -2$

14	$\dfrac{1}{8} < m < 8$ *(number line from −3 to 11, open circle at 0, open circle at 8, shaded between)*
15	$a \le \dfrac{1}{3}$ *or* $a > \dfrac{39}{10}$ *(number line from −4 to 8, closed circle near 0, open circle near 4, shaded left and right)*
16	$7 \le n \le \dfrac{50}{7}$ *(number line from 2 to 14, closed point at 7)*
17	$-1 \le x < 6$ *(number line from −4 to 9, closed circle at −1, open circle at 6, shaded between)*
18	$-2 \le m < \dfrac{48}{7}$ *(number line from −5 to 7, closed circle at −2, open circle at 7, shaded between)*
19	$-\dfrac{24}{7} < v < \dfrac{21}{4}$ *(number line from −6 to 6, open circle near −3, open circle near 5, shaded between)*
20	$x \ge \dfrac{5}{3}$ *or* $x \le \dfrac{1}{2}$ *(number line from −5 to 7, closed circle near 0, closed circle near 2, shaded left and right)*

Graphing

1	A number line from −7 to 7 with a closed circle at −2 and the ray shaded to the right.
2	A number line from −7 to 7 with a closed circle at 3 and the ray shaded to the left.
3	A number line from −7 to 7 with a closed circle at −2 and the ray shaded to the right.
4	A number line from −7 to 7 with a closed circle at 6 and the ray shaded to the left.
5	A number line from −7 to 7 with an open circle at 0 and the ray shaded to the right.
6	A number line from −7 to 7 with an open circle at −6 and the ray shaded to the right.
7	A number line from −7 to 7 with an open circle at 5 and the ray shaded to the left.
8	A number line from −7 to 7 with a closed circle at 4 and the ray shaded to the left.
9	A number line from −7 to 7 with an open circle at 6 and the ray shaded to the left.
10	A number line from −7 to 7 with an open circle at −2 and the ray shaded to the left.

11	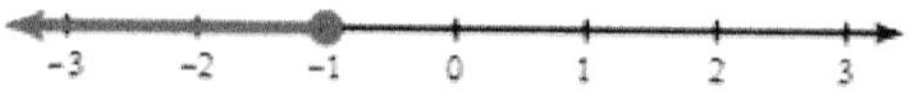
12	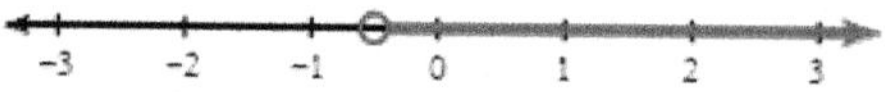
13	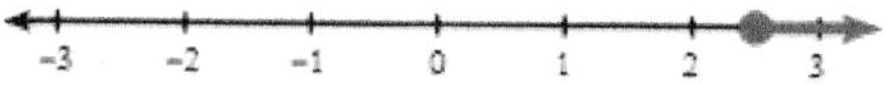
14	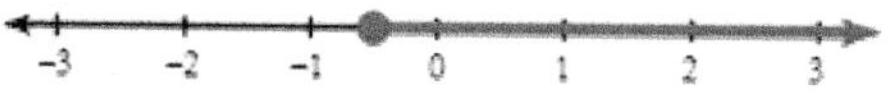
15	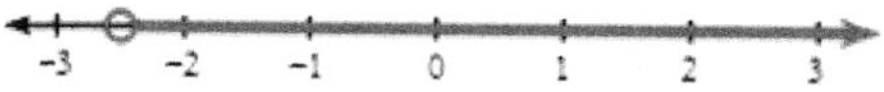
16	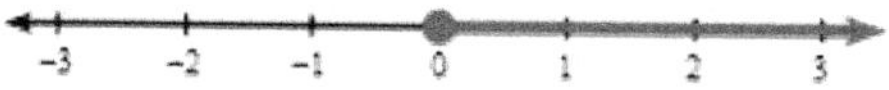
17	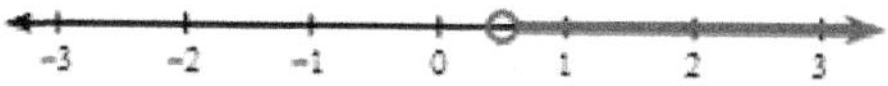
18	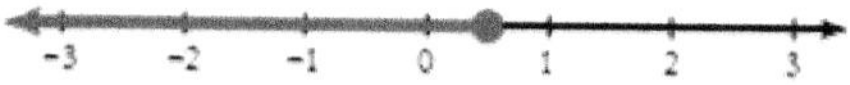
19	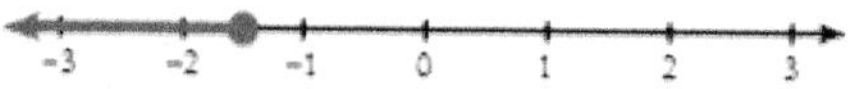
20	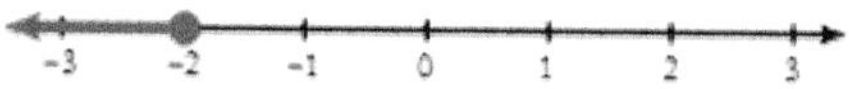

One Step Inequalities

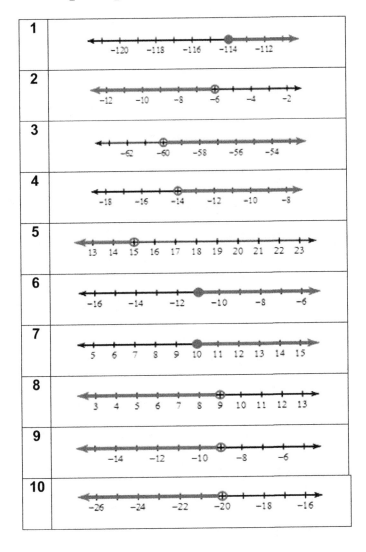

11	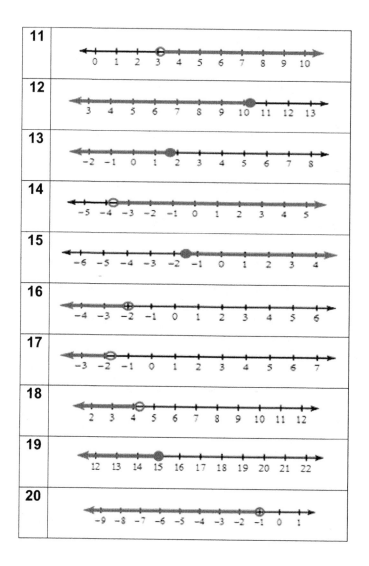
12	
13	
14	
15	
16	
17	
18	
19	
20	

Two Step Inequalities

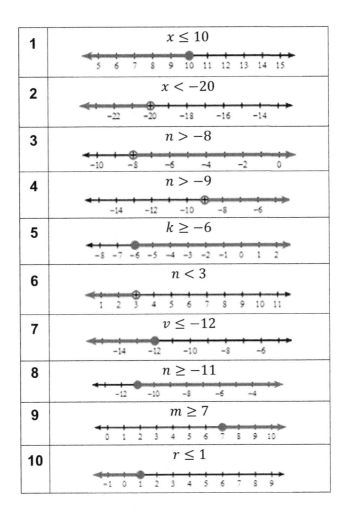

1	$x \leq 10$
2	$x < -20$
3	$n > -8$
4	$n > -9$
5	$k \geq -6$
6	$n < 3$
7	$v \leq -12$
8	$n \geq -11$
9	$m \geq 7$
10	$r \leq 1$

12	$v < -\dfrac{35}{2}$ (number line from −24 to −14, open circle at −17 shaded left)
13	$a < \dfrac{119}{12}$ (number line from 7 to 17, open circle at 10 shaded left)
14	$k \leq \dfrac{4}{5}$ (number line from −3 to 7, closed circle at 1 shaded left)
15	$x < \dfrac{13}{2}$ (number line from −1 to 9, open circle at 7 shaded left)
16	$x \leq \dfrac{139}{16}$ (number line from 2 to 12, closed circle at 9 shaded left)
17	$n \leq \dfrac{129}{14}$ (number line from 5 to 15, closed circle at 9 shaded left)
18	$m > \dfrac{47}{10}$ (number line from 2 to 12, open circle at 5 shaded right)
19	$x \geq -\dfrac{5}{6}$ (number line from −8 to 2, closed circle at −1 shaded right)
20	$n < -\dfrac{12}{13}$ (number line from −4 to 6, open circle at −1 shaded left)

Multi Step Inequalities

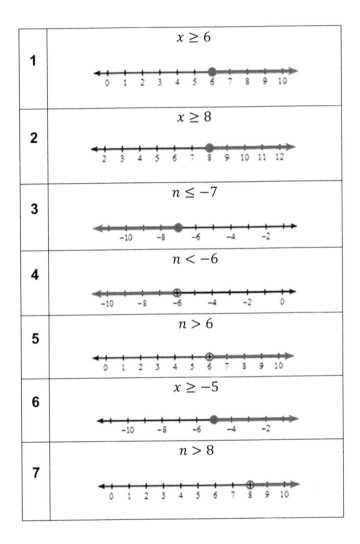

1	$x \geq 6$
2	$x \geq 8$
3	$n \leq -7$
4	$n < -6$
5	$n > 6$
6	$x \geq -5$
7	$n > 8$

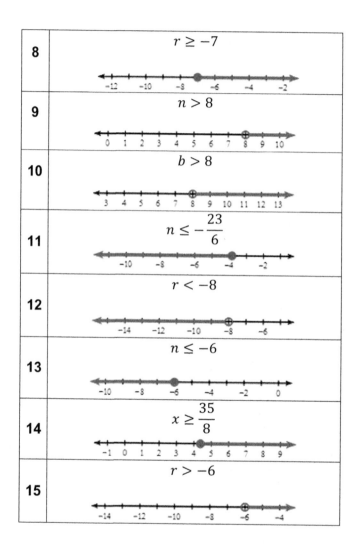

8	$r \geq -7$
9	$n > 8$
10	$b > 8$
11	$n \leq -\dfrac{23}{6}$
12	$r < -8$
13	$n \leq -6$
14	$x \geq \dfrac{35}{8}$
15	$r > -6$

16	$r \geq 7$ 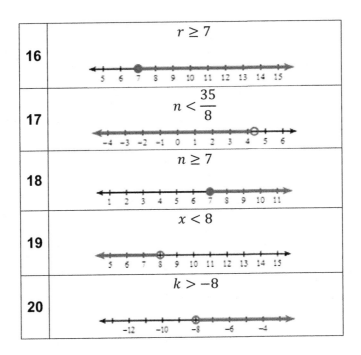
17	$n < \dfrac{35}{8}$
18	$n \geq 7$
19	$x < 8$
20	$k > -8$

Graph the solution

1	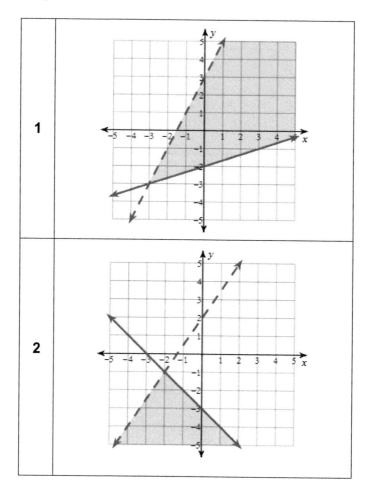
2	

3	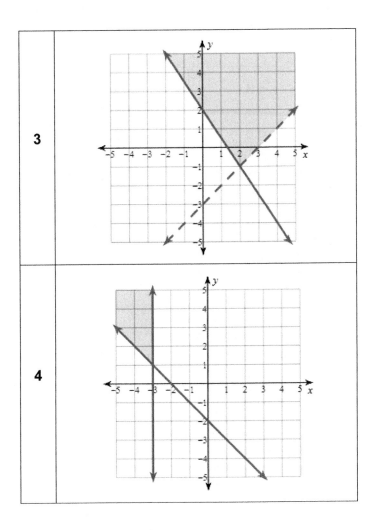
4	

Writing given a graph

1	$x < 3$	**11**	$x \leq 4$
2	$n < 0$	**12**	$k \leq 3$
3	$x > 3$	**13**	$x > 1$
4	$v \geq 0$	**14**	$x \geq -6$
5	$x \leq -5$	**15**	$x \leq 6$
6	$n < -5$	**16**	$a < -3$
7	$n < -2$	**17**	$x > -1$
8	$a \geq -4$	**18**	$n \geq 3$
9	$x \leq 5$	**19**	$k \leq 2$
10	$b \leq 0$	**20**	$x \leq -4$

LINEAR FUNCTION

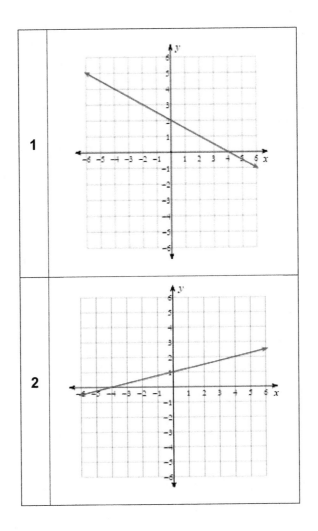

| 3 | 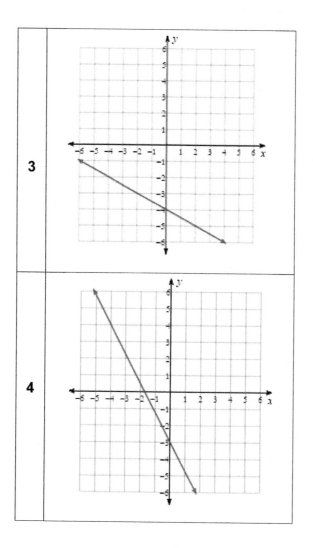 |
| 4 | |

SLOPE OF A LINEAR FUNCTION

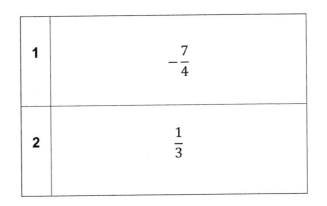

1	$-\dfrac{7}{4}$
2	$\dfrac{1}{3}$

LINEAR INEQUALITIES

Graphing Linear

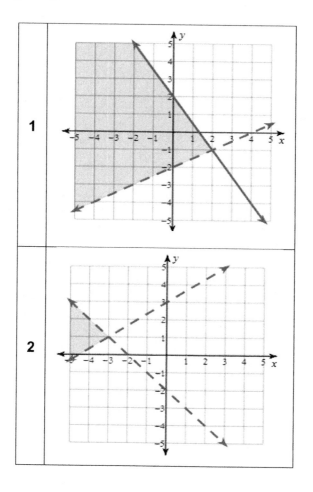

3	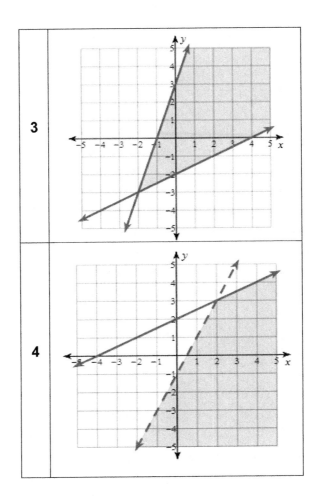
4	

Graphing Systems of Inequalities

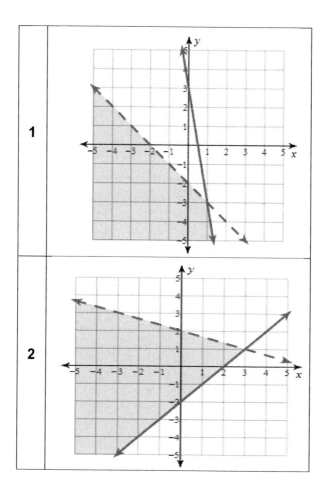

3	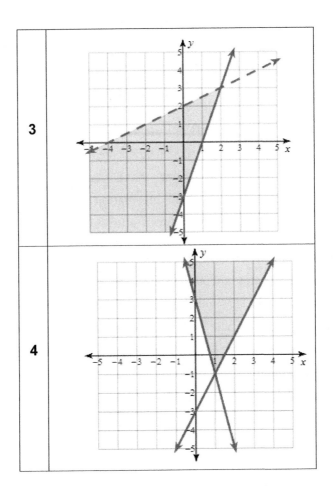
4	

Conclusion

Thank you for making it through to the end of *Algebra 1 Workbook: The Self-Teaching Guide and Practice Workbook with Problems and Related Explained Solution. You Will Get and Improve Your Algebra 1 Skills and Knowledge from A to Z.* Let's hope that this book has provided you with some useful information on this topic that will be helpful to you in your future study of mathematics.

The next step for you to complete here would be to further practice the topics laid out here until you have developed an algebraic vocabulary and skill sufficient to solve any future problems that you will be beset with. Thank you again for downloading this Book. Hopefully, it was a helpful one while still being perhaps less dry than most algebra books on the market.

Made in the USA
Middletown, DE
04 May 2021